Studies in

American Popular History and Culture

Edited by

Jerome Nadelhaft

University of Maine

A Routledge Series

Studies in American Popular History and Culture

Jerome Nadelhaft, *General Editor*

Writing Jazz
Race, Nationalism, and Modern Culture in the 1920s
Nicholas M. Evans

Automobility
Social Changes in the American South, 1909–1939
Corey T. Lesseig

Actors and Activists
Politics, Performance, and Exchange Among Social Worlds
David A. Schlossman

Studies in the Land
The Northeast Corner
David C. Smith

First Do No Harm
Empathy and the Writing of Medical Journal Articles
Mary E. Knatterud

Piety and Power
Gender and Religious Culture in the American Colonies, 1630–1700
Leslie Lindenauer

Race-ing Masculinity
Identity in Contemporary U.S. Men's Writing
John Christopher Cunningham

Crime and the Nation
Prison Reform and Popular Fiction in Philadelphia, 1786–1800
Peter Okun

Food in Film
A Culinary Performance of Communication
Jane Ferry

Deconstructing Post-WWII New York City
The Literature, Art, Jazz, and Architecture of an Emerging Global Capital
Robert Bennett

Rethinking the Red Scare
The Lusk Committee and New York's Crusade against Radicalism, 1919–1923
Todd J. Pfannestiel

Hollywood and the Rise of Physical Culture
Heather Addison

Homelessness in American Literature
Romanticism, Realism, and Testimony
John Allen

No Way of Knowing
Crime, Urban Legends, and the Internet
Pamela Donovan

The Making of the Primitive Baptists
A Cultural and Intellectual History of the Antimission Movement, 1800–1840
James R. Mathis

Women and Comedy in Solo Performance
Phyllis Diller, Lily Tomlin, and Roseanne
Suzanne Lavin

The Literature of Immigration and Racial Formation
Becoming White, Becoming Other, Becoming American in the Late Progressive Era
Linda Joyce Brown

Popular Culture and the Enduring Myth of Chicago, 1871–1968
Lisa Krissoff Boehm

AMERICA'S FIGHT OVER WATER

The Environmental and Political Effects of Large-Scale Water Systems

Kevin Wehr

Routledge
Taylor & Francis Group
NEW YORK AND LONDON

Published in 2004 by
Routledge
711 Third Avenue, New York, NY 10017

Published in Great Britain by
Routledge
2 Park Square, Milton Park, Abingdon, Oxfordshire OX14 4RN

Routledge is an imprint of the Taylor & Francis Group.

Transferred to Digital Printing 2008

First issued in paperback 2012

Library of Congress Cataloging-in-Publication Data

Wehr, Kevin, 1972–
America's fight over water : the environmental and political consequences of large-scale water systems / Kevin Wehr.
p. cm. — (American popular history & culture)
Includes bibliographical references and index.
ISBN 0-415-94930-0 (hardback : alk. paper)
1. Water resources development—Political aspects—West (U.S.)—Case studies. 2. Water resources development—Social aspects—West (U.S.)—Case studies. 3. Water resources development—Environmental aspects—West (U.S.)—Case studies. 4. Glen Canyon Dam (Ariz.) 5. Hoover Dam (Ariz. and Nev.) 6. Grand Coulee Dam (Wash.) I. Title. II. Series: American popular history and culture (Routledge (Firm)).
HD1695.W4W38 2004
333.91'62'0979—dc22 2004005112

ISBN 13: 978-0-415-94930-9 (hbk)
ISBN 13: 978-0-415-64580-5 (pbk)

Why should you not spend money to . . . defend your government, your society, against your expensive paupers and criminals? We must show the poor man how he can afford to get married . . . and become a homeowner.

General Frederick Booth-Tucker of the Salvation Army, 1895

The superiority of the western half-continent over its eastern counterpart may not be expressed in a word. It is, rather, a matter for patient unfolding through a study of natural conditions over wide areas, and a scrutiny of the human institutions which are the inevitable product of this environment. Aridity, in the elementary sense, is purely an affair of climate. That it is also the germ of new industrial and social systems, with far-reaching possibilities in the fields of ethics and politics, will be demonstrated.

William Ellsworth Smythe, 1899

The farm boys in the East want farms of their own. [The Newlands Act] gives them a place where they can go and build homes without being driven into the already crowded cities to seek unemployment.

Representative Oscar Underwood of Alabama, 1901

In the shadow of Hoover Dam one feels that the future is limitless, that no obstacle is insurmountable, that we have in our grasp the power to achieve anything if we can but summon the will.

Joseph Stevens, 1988

It required 19 centuries to define decent man-to-man conduct and the process is only half done; it may take as long to evolve a code of man-to-land conduct.

Aldo Leopold, 1949

What we know now that we didn't know in 1938 is that a river isn't a water pipe.

Bruce Barcott, 1999

Contents

Preface ix

Chapter One
Introduction: The Fight over Water in the American West 1

Chapter Two
Political and Environmental Sociology:
The Dialectic of Society and Nature 9

Chapter Three
Water in the American West 41

Chapter Four
In the Beginning There Was Boulder:
A Natural Menace Becomes a Natural Resource 59

Chapter Five
Grand Coulee: "Mightiest Thing Ever Built by Man" 135

Chapter Six
Glen Canyon: Last of the High Dams 187

Chapter Seven
DamNation: Controlling the Waters, Civilizing the Wilderness 225

Maps 239

Notes 243

Bibliography 251

Index 265

Preface

America's fight over water has long historical roots: from indigenous communities clustered near year-round flows to the twentieth century Anglo domination of the wild rivers of the West, water has been central to the social, political, and cultural life of the western portions of North America. Comparative historical sociology can help make sense of socio-cultural shifts over time, and this work addresses how the control of nature has shifted from sustainable use to domination, and how the unintended consequences of this history pose serious problems for sustainability in the future.

The central organizing principle of the book is an examination of the discourses about the proposal and construction of large-scale water systems in the American West. The fights over water stretched from the shifting northern borders with Canada to the contested southern borders with Mexico, and from the emptying of the rivers into the Pacific Ocean to the corridors of power in Washington, D.C. Multiple social groups argued for and against construction of the dams, hydropower plants, and irrigation canals, and the rhetoric these groups used betrays a complexity of continuity with past views of nature's gifts to society and changing views of how humans should preserve natural resources. The three case studies presented—Boulder (Hoover) Dam, Grand Coulee Dam, and Glen Canyon Dam—exhibit these complexities, as well as the power and politics of the social groups competing for dominance.

The examination of these three high dams in the American West shows how the discussions around the dams illuminate a particular set of society-nature relationships. Society and nature interact with—and mutually construct—one another in complex ways, but an approach that utilizes the perspectives of political and environmental sociology can yield important insights regarding the ways that culture, politics, and nature are linked. Nature and society exist in a dialectical relationship, where the growth of society is influenced by natural barriers, and nature is concomitantly effected by social forces. Building dams in an arid land is one way to overcome natural barriers to social development, but stopping the natural flow of a river is not just costly, it has clear environmental consequences. Such costs and consequences need to be justified. This book addresses the specific question: what methods were used to justify and/or contest the building of an economic and political infrastructure in response to the perceived water scarcity of the American West?

To address this question, I examine the historical discourse around three dams, focusing on how they were presented to the public by the state and how the public received them. The physical existence of the dams has no *inherent* meaning; rather, different social groups assign meaning to the dams. My central focus, then, is on how the representations of the dams were mobilized in order to legitimate an ideology based in the domination of nature. How was this ideology supported, mystified, and contested? Examining the material culture of the dams (the publications, photos, brochures and other forms of discourse used to justify their proposal) helps to show how a dominant ideology can be translated into material reality. I argue that the ways the dams were represented—how the state, private capital, and local inhabitants in the West viewed the potential of the dams, and later how they viewed the various intended and unintended consequences—legitimated an ideology of unilinear progress through the application of science and technology to overcome perceived natural problems. I believe that the views of the dams shows an acceptance of this dominant ideology of subjugating nature; the representations of the dams and of nature and society function to legitimate state building in the form of human transformations of the landscape.

This work shows that the discourses surrounding the proposal and construction of large-scale water systems are real—they are not just texts to be read, interpreted, and deconstructed. The ways that people and institutions construct reality have real effects on the landscape of the West. Indeed, they have a real impact on people's lives, their livelihood, and even their bodies. The justifications that enabled the building of the dams helped to produce the metropoles of Los Angeles and Las Vegas, Seattle and Portland. They helped California and the West grow to prominence in the United States' economics, politics, and culture. We have stepped onto a treadmill of development, and once on, it may be extremely difficult to get off.

Without water and hydropower, the United States would have a very different west coast, one with a much more diminutive population. With such a demographic shift the national political landscape would have an altered topography. Water politics in the West thus holds an important place in U.S.

history for many reasons. Modernist discourses legitimating state building and socio-economic expansion helped to bring this about, and environmentalist discourses helped to put the brakes on this blind development.

A new historical analysis of state building rhetoric and the fight over water in the American West is important for other reasons as well. Though the problems with such development strategies are reasonably well known, there are currently construction plans for high dams in many nations of the global South. Political regimes as varied as those of Chile, India, and China are all in the process of building dams for all of the same reasons that the technology was first created in the western U.S. These nations all use similar rhetoric as the New Deal politicians and dam boosters, and appear to ignore the unintended consequences that came with the dams of the American West. How is it that the same rhetoric is used in cases separated by 100 years and several continents? Will the outcomes be the same? A deep examination of the first place and time when such projects were justified may help us understand the reasons and effects of their contemporary mobilization.

Any analysis that reaches toward objectivity on a controversial issue must put forward the author's biases at the outset. I declare myself ambivalent (in its denotative sense: I am of two minds) on the issues of water and the West. Some environmentalists portray the big dams as the product of villains and as the physical embodiment of evil—they fantasize along with Ed Abbey about blowing the dams sky-high. While I sympathize with this fantasy I can't say that I embrace it, for my life—as was Abbey's—is materially dependent upon those big dams. My great-great-great grandfather moved to California in 1848, seeking gold. The other side of my family moved to the West Coast after World War II, settling in well-watered northern California. This lineage represents two extremes of Anglo California inhabitants: the rugged individual encountering the "wilderness" of the frontier and the middle class baby-boomer family enjoying post-war prosperity.

So while my radical-environmentalist side cherishes the Earth First! dream of breaching Glen Canyon, the pragmatic side realizes what the damming of the West has wrought: an exploding economy, a booming population, and enough hydro-electric power to help win World War II and propel the United States to a hegemonic position in the global economy, with all of the minutia of all the lives that depend upon all of this. Clearly these are not unmitigated positives, but some aspects have enough worth to convince me to maintain a healthy ambivalence and skepticism for those who speak in superlatives—both for and against the dams. It is my hope that this book will help to make sense of the controversy and discordant perspectives that swirl around the dams of the American West.

The usual bromides apply to this book as to so many others. While responsibility for the work is mine alone, without the help of so many people it would not have happened. I would especially like to thank Jess Gilbert for his indispensable advice and guidance, as well as the support and encouragement of my other mentors: Fred Buttel, Jane Collins, Bill Cronon, and Phil Gorski, all of whom have been both supportive and encouraging. Hundreds of librarians and archivists helped along the way, and I greatly appreciate their assistance in uncovering manuscripts and images (every effort has been made to locate the copyright holders, the works have been properly cited throughout, and I appreciate permission for reprinting material). I have to particularly acknowledge the forbearance of my writing group, Alisa Rosenthal and Jeff Helmers, who tolerated my at times spasmodic attempts at writing and rewriting to make it better than good. Thank you also to Susan and Jason; Martha, Matt, Monica, and the union and CD crew; Audra, Becky and the co-opers, and of course Jacquie for all of your support. This book is dedicated to Sadie, the one who carried me the whole way.

Kevin Wehr
Sacramento

Chapter One

Introduction: The Fight over Water in the American West

Oh the world is seven wonders
so the travelers always tell.
Some gardens and some towers
I guess you know them well.
But now the greatest wonder
is in Uncle Sam's fair land.
It's the king Columbia River
and the great Grand Coulee Dam

"Grand Coulee Dam" Woody Guthrie

"Water, water, everywhere / Nor any drop to drink."

Samuel Taylor Coleridge

Modern poets, such as Woody Guthrie, have lauded the achievements of the newest world wonders—the dams of the American West. This myth-making is central to the success of the dams and the building of an ideology regarding nature. But modern poets, whose works form a thread throughout this book, have not recognized what Samuel Taylor Coleridge alludes to in "The Rime of the Ancient Mariner." Coleridge's thirsty sailor is ironically surrounded by undrinkable salt water. This eighteenth-century poem may soon have some contemporary relevance in the American West. As demand for water increases, dams are silting up and supply is dwindling. Agribusiness intensively irrigates mono-crop fields, producing mineral-rich run-off that makes rivers saltier than seawater. Meanwhile, more and more people move to Western states, and federal policy has shifted, curtailing the state's ability to respond to a looming water crisis.

What hubris led humans to the brink of this potential tragedy? The American West is not so different from other societies in its profound alterations of its natural environment—what Paul Jacobson calls "brute force technology" (2002: 4–8)—and yet the scale, intensity, and consequences make the West an interesting case: rarely have changes to the environment been so immense and numerous as the dams of the American West. Large-scale water systems have been built on all major rivers of the West; dams trap water for drinking and irrigation, provide hydroelectric power, control flooding, and ease navigation. The dams are also among the premier stops as tourists experience the grandeur of the West, and the massive reservoirs have become a playground for anglers, water skiers, and campers.

Most of the dams in the West were built during the Progressive Era and the New Deal, and consequently reflect an ideology of rational planning and state building based in a faith in scientific progress. State-sponsored infrastructure had myriad environmental and political effects, but the natural formations[1] that the state worked to overcome also had a profound influence on how society developed. Through an examination of the ways that nature, society, and the state interact with and mutually construct one another, this work will contribute to an integration of political and environmental sociology.

The theoretical impetus for this effort is to illuminate the relationship between society and nature. This theoretical concern is two-fold. First, political and environmental sociology, respectively, have contributed important insights towards understanding the ways that culture and politics are linked and the ways that society and nature are linked, but rarely are these areas integrated. Through an examination of dams in the American West, this work makes sense of the ways that central concerns of political sociology—state-building and social movements—influence and are influenced by the nature-society relationship, which is a central concern of environmental sociology. The state does not just build an infrastructure on a passive landscape. Rather, a conception of the relationship between nature and society must be discursively constructed, in part through bottom-up cultural processes as well as state-directed top-down ideological processes. Second, there were specific social and environmental effects of this state-building process that contributed to a nature-society relationship that has dialectically changed over time. Since the natural environment is not simply a passive object that the state builds upon, the analysis of discourses, manifested in cultural texts, can help integrate political and environmental sociology by contributing to the understanding of the ways that natural conditions helped and/or hindered state building. This book thus asks the specific question: what discursive methods were used to justify or contest the building of an economic and political infrastructure in response to the perceived water scarcity of the American West?

Recent political sociology has taken a "cultural turn" (Steinmetz 1999) towards a broadened interest in the constitutive role of culture in politics and the state. Discourse analysis has played a large part of this turn; by examining how meaning is constructed, maintained, and transformed, scholars have helped to illuminate the ways that the political realm is constituted, in part, by culture. To fully understand the social world we must not only examine political-economy, but also cultural, spatial, and temporal forces. This work builds on the cultural turn in the discipline of sociology by endeavoring to connect the discourses about nature and state building on the landscape of the West—a subject that is both political and environmental. The connections between politics and environment abound, and discourse analysis can help clarify and explain those connections.

To address these connections, this book will examine the discourse around the dams of the American West: how they were presented to the public by the state and how the public received them. The physical existence of the dams has no inherent meaning; rather, different social groups assign meaning to the dams. My central question, then, is how were the representations of the dams mobilized—supported, mystified, and contested—through a discourse of high modernism[2] based in the domination of nature? Examining artifacts of the material culture of the dams (the publications, photos, brochures, and other forms of discourse used to justify the construction of the dams) will help to show how a dominant ideology can be translated—albeit unevenly—into materiality. In the spirit of Chandra Mukerji's (1997) work on the ways that the gardens of Versailles represented control over France's territory, I will show the ways that the dams of the American West symbolize the human transformations of the landscape under high modernism.

To better illustrate the issues at the heart of this study, I have singled out a few dams to explore in greater detail. But the larger argument I make is that the discursive representation of the dams and the relationship between nature and society operate for all of the dams in the West, whether they were built by private capital, individual states, the Army Corps of Engineers, or the Bureau of Reclamation. Avoiding a singular, place-specific case study, I instead explore the ways that the general discourse around dams was created and consumed throughout the nation, both on a local level, within debates in Congress, and in the national popular media. The several dams I examine closely are Boulder Dam on the lower Colorado River (chapter four), Grand Coulee Dam on the Columbia River (chapter five), and Glen Canyon Dam on the upper Colorado River (chapter six), which are all represented in the map section, pages 239–41.

This work, then, examines the role of discourse in legitimating the construction of large-scale water systems across the western half of the United States. The dams represent material power—the infrastructural expansion of the state, the exertion of state control over territory, and the domination of nature. The discursive practices and ideologies that legitimated the building of the dams show that the dams represent a prime example of high modernism in its monolithic pretensions and contested results.

I argue that the ways that the early dams (Boulder and Grand Coulee) were represented—how the state, private capital, and local inhabitants in the West viewed the potential of the dams, and later how they viewed the various intended and unintended consequences—legitimated an ideology of high modernism. This ideology centered on a process of state building that transformed the landscape of the West to enable political, economic, and social control and to spur agricultural and industrial development. Such transformations represent an example of a "spatial fix" that capitalism must achieve in order to stave off the continual threat of crisis (Harvey 1982, 2000: 23–30).

As with any hegemonic ideology, there is some amount of contestation apparent in the debates over the dams. The opposition, however, was rooted in economic self-interest. Detractors of the early dams were concerned with the *type* of dam that would be built—and specifically who would benefit—rather than *whether or not* the dams should be built (chapters four and five). This changed over the course of the mid-twentieth century. With the rise of an oppositional discourse rooted in an ideology different from high modernism, opponents could effectively contest the dams at Echo Park and Glen Canyon in the Colorado River Basin (chapter six). This shift towards an effective oppositional discourse posed an important challenge to the hegemony of the high modernist ideology. The absorption of this oppositional discourse into high modernism has proved to be uneven (chapter seven).

We can identify three levels, or layers, of discourse production: the actual production of the images or texts, the significance of the intended consequences, and the consumption of those images and texts, including any unintended consequences (Lutz and Collins 1993). The first level of discourse is represented by the texts produced by the U. S. Department of Interior's Bureau of Reclamation. There are three broad levels of Bureau documents: general publications about the need for (and the glory of) the dams, reports about particular dams and how they are important for the local area and the region (or the nation as a whole), and detailed engineering documents about the construction and management of particular dams. These documents form a central core of data for this study, with other groups and individuals responding to or initiating Bureau actions. I compare these representations of the dams with debates in Congress, oppositional voices, local and regional boosters, and the national press.

Thus, the discourses regarding the large-scale water systems of the American West were produced by situated subjects within an ideological arena. As the discourses about the environment of the West proliferated around the turn of the twentieth century, views of nature and society came into conflict. Different social groups mobilized ideas and images to support or attack various forms of intervention in natural processes. Discussions regarding the virtues of building large-scale water systems provide a window into the views of nature.

The discourse around the dams—regarding the construction, management, how they were "sold" to the public, and how the public received them—is a multi-vocal subject. Many social groups contributed to the discourse of high modernism around the dams: the state (which is not a unitary object: this includes Congress, the President, the different Bureaus and agencies), displaced groups, grassroots organizations, farmers, urban consumers of water and agricultural produce, and tourists. This discourse is totalizing and imperialistic, both in its ideal type and as it was exemplified by dam boosters and detractors in the American West. The definition of the ideology of high modernism would be incomplete without showing the ways that discourses of resistance are co-opted and incorporated into the dominant discourse of high modernism. High modernism assumes contestation, but also often has the ability to absorb this resistance of disparate voices into an uneasy—and often uneven—unity.

THE STRUCTURE OF THE COMING CHAPTERS

This book is easily divided into two (unequal) parts. The first three chapters offer a broad overview of the historical trajectory of dam building and state intervention in the West. I first suggest, define, and discuss some theoretical concepts that will guide the analysis of the empirical cases. I follow these sections with a critique of the division of nature and society into two separate analytical subjects. This discussion will take place alongside a truncated examination of the history of the American West. Readers who are more interested in the history and empirical analysis of the case studies may wish to skip the theoretical and historiographical discussions of chapters two and three and flip directly to chapters four through six.

Environmental sociology's analytical division of society and nature (industry and the polluted environment, farm and agricultural market, city and country) is, I argue, unfounded when compared to the historical record.

Instead, following David Harvey (1996), I take a dialectical approach regarding the society-nature relationship. Such an approach allows the complexity of the relationships between the (constructed) concepts of society and nature to be explored more fully. An examination of the larger historical literature on the environment of the American West supports this approach, as will be shown in chapter two.

The discipline of environmental history has long been concerned with nature-society relationships, and the historical record shows the complex and subtly changing human relationship to the natural environment. The increased attention to environmental history in the last 20 years has shown the profound alteration of nature by humans. In the American West in particular the changes wrought by humans have been significant; and yet the West has been the subject of many myths, and historians of the American West have shown how the construction of myths supporting human changes in the land have been central to everyday existence (see, for example, Cronon, Miles, and Gitlin 1992). These myths of the West are woven through discourse: the stories of brave trappers battling Indians, tales of pioneers entering a chosen land of plenty, or even folk singers praising the new world wonder of "the great Grand Coulee Dam," as Woody Guthrie did when he was hired to write songs for the Department of the Interior. In essence, Guthrie created myths about the Columbia River—or at least elaborated and expanded on the myths that were nascent in the Pacific Northwest. Environmental and political sociology have much to learn from the insights of recent historical work on the interplay of history, nature, and culture. The next two chapters of this book attempt to include and build upon such insights.

The project then moves to the empirical cases, which make up the bulk of the work. Chapters four through six consist of deep examinations of the discourses around the three dams. Each chapter examines an individual dam, chosen for its symbolic weight, its importance to the West, and variation around booster and oppositional groups. Within each chapter, I will examine the logic of the state, the Bureau of Reclamation, private capital, and local supportive and oppositional groups in relation to each dam project. This method of organizing the work allows for polyvalence within each project, where both contestation and broad agreement on issues can be underscored. Continuities and differences between projects will be highlighted and analyzed as the work proceeds.

I have chosen dams as the case studies for this project because water scarcity defines the American West (Webb 1931/1959: 17). From the 98th Meridian to the Pacific Ocean, annual average rainfall is less than twenty inches, except for some areas of the Pacific Northwest and the mountain regions.[3] Over the last century in this dry land, population increases and economic growth have been fueled by the parallel processes of the conversion of public land to private ownership, the concentration of capital, and the control of labor. These political-economic forces, in turn, depend upon a ready supply of water.

The U. S. government recognized water scarcity as an important issue near the end of the nineteenth century. The construction of hundreds of dams under the auspices of the Bureau of Reclamation, the Army Corps of Engineers, and individual states formed an infrastructure that, in a circular and cumulative manner, encouraged dramatic increases in population, economic might, and political power. These developments also had major effects upon the environment. From the obvious changes represented by the creation of canyon-filling reservoirs to the more subtle increases in soil salinity from intensive irrigation, the large-scale water systems of the American West have had a profound influence on environment, economy, and society. But the influence is dialectical. The actions of the state, in the attempt to surmount the barriers of an arid land, were constrained by existing natural conditions and nature's response to anthropogenic environmental changes.

The empirical chapters of this book address the complexity of this dialectical interplay between state, culture, and environment. The first two empirical chapters examine Boulder and Grand Coulee Dams; two early dams that saw no opposition other than differences in opinion regarding the form of the dam. Using imperialistic rhetoric, boosters talked about the need to build an empire in the region, or argued that the dam would contribute to the economic progress and growth of the region. At Grand Coulee this was inflected by either a high modernist, state-centered discourse or by an individualist-capitalist discourse that focused on private capital rather than the state. Hebraic discourses focused on the people who lived in the region, presenting them as a "chosen people" who lived in a Land of Canaan. Such a Promised Land should, they argued, be completed by a series of state-built dams.[4]

These discourses at Boulder and Grand Coulee show how that the lack of resistance to high modernism enabled continued hegemony. The discourses around the early dams all centered on a domination of nature that was not contested. At Glen Canyon Dam, however, this discourse was strongly challenged. The rise of an environmental discourse allowed an effective opposition to new dams in the American Southwest in the 1950s. This rediscovery of the preservationist discourse of John Muir was used to combat the hegemonic ideology of high modernism, which was triumphant in the American West from the building of Boulder Dam through the 1950s.

The Glen Canyon chapter explores the rise of this discourse, its consequences, and the cooptation that it has experienced.

The ultimate goal of the work is to provide an understanding of the ways that historical discourse has shaped the nature-society dialectic and how state building is different in the American West because of this. This goal will be most concretely addressed in the concluding chapter.

Chapter Two

Political and Environmental Sociology: The Dialectic of Society and Nature

> This land is your land, this land is my land
> From California, to the New York island
> From the redwood forest, to the Gulf Stream waters
> This land is made for you and me
>
> "This Land Is Your Land" Woody Guthrie

Although the dams of the American West loom large over society, much of their political character is hidden from view. Snippets of politics come back into the media now and again; talk about breaching the dams rises and falls with the issue-attention cycles of the American public (Downs 1998). The history of the dams, however, almost always remains buried underneath the present discussions about endangered species or flood control or urban water needs.

Much of political sociology has been concerned with exposing such hidden transcripts, and this work will examine three case studies of the control of nature in the American West with an eye towards understanding the complex interactions of politics, nature, and society. Attempts to resolve this complex set of interactions, however, have tended towards easily described models: a binary of base-and-superstructure or state-and-society, or stratified hierarchies of influence. These metaphors do not accurately portray the complexity of the political world or the natural world; instead, the concept of the dialectic, as recently described by David Harvey, offers a better foundation for this analysis. Society and politics and environment are thoroughly imbricated in one another and cannot easily be separated, even for analytic purposes. Theoretical models and metaphors, of course, inherently reduce the complexity of reality so as to understand the fundamental causes and significant characteristics of society—both past and present. This chapter looks to the idea of the dialectic to avoid undue reductionism and to respect the complex and contingent ways that society and nature interrelate.

Similar to Woody Guthrie's deceptively simple lyrics of "This Land Is Your Land," the relationship between nature and society appears transparent, but actually obscures property relations, bounded territories, and assumptions about citizenship and statehood. The relationship between nature and society is certainly complex, but the fundamental characterization of this relationship remains that of domination. The chapter thus begins with an examination of the terms discourse and ideology, and moves to a definition of the concept of domination, including its gendered character. This is followed by a brief discussion of alternative relationships between society and nature throughout history, and how domination came to be primary in modernity. Furthermore, the state, civil society, culture, history, and geographic space interrelate in complex ways that cannot be easily understood within the confines of a single discipline. Building on insights from history, geography, and environmental and political sociology, this chapter argues that discourse about social relationships with nature is an important addition to understanding how state building works, particularly in the American West.

DISCOURSE AND IDEOLOGY

What is discourse, and how does it change over time? Examining historical discourses to illuminate the relations between society and the environment implies both a theory of history and a theory of knowledge. History, as understood for this work, is not an abstract agglomeration of facts and contingencies, but rather can be seen as simultaneously chaotic and patterned—yet teleological theories have little applicability in modern social research. Instead, the trick of historical sociology is to sift through the details of history to examine the patterns induced by the interaction of individual agents, social groups, social structures, and, in the cases presented here, the natural environment. This eliding of human and environmental history helps destabilize the anthropocentric aspects of many historical accounts, and shows the patterns of history as they are expressed upon the landscape. The rivers of the West offer an opportunity to read the history of human interpretations of nature: rivers may be viewed as life-giving resources, the embodiment of an aesthetic sublime, or as an "organic machine"[1] capable of providing a dense population with its energy, sustenance, transportation, and recreational needs.

The examination of historical discourses can similarly show the patterns of the history of human thought about the relations between society and the natural environment.[2] To maintain a historical and material grounding, such an examination must be carried out with conscious reference to political, economic, and social contexts. For this study, such grounding means an emphasis on complex power relations—including political, economic, military, and ideological power (Mann 1990)—particularly those ideologies that drive humans to dominate nature to gain economic and political ends. These power relations are not only expressed in discourse, but also in the consequences of discourse: material culture such as the towering dams, tentacle-like reservoirs, and snaking canals of the large-scale water systems, the abundant mono-crop fields, and the ever-multiplying suburban housing tracts that support a political system that remains beholden to such constituents. It is crucial to emphasize, as Harvey (1996) has argued, that ideas matter when they are connected to systems of power and are thus expressed in a material way.

Similar to the analysis of history, examining discourse implies an epistemology and ontology. This work assumes that, although social perceptions and the naming of reality involve important cognitive processes that condition the perception of reality (and have concrete effects, *viz.* Berger and Luckman [1967]), I argue that reality is far from wholly socially constructed. Nature, to put it bluntly, also shapes human society and history. There are certain patterns that nature operates within and this offers a material basis for what is sometimes construed as a wholly socially constructed reality. Society, however, also "produces nature" (Smith 1984) to the extent that some "universal" laws can be manipulated in order to benefit humans (within certain bounds). The laws of evolution may apply to all species, but, since the tool-making age, hominids have had an active role in directing their own and other species' evolution through the use of tools (Haraway 1978). The recent advances in biotechnology certainly raise questions of a radical new ability to direct human and plant evolution. Human-directed evolution has actually gone on for many thousands of years through animal and plant husbandry, plant cross-breeding, as well as the pernicious practices of genocide, slavery, and miscegenation laws. There are, then, some "laws" of nature that "govern" both human and natural reality.

One of the sources of the debate regarding the social construction of reality lays in the connection between discourse, ideology, and their material consequences. Humans construct reality, it is argued, through filtering perceptions, habituated ways of thinking, and the act of labeling: all processes that relate to discourse and occur within ideological fields. The ideas of Noam Chomsky (1957) and Stephen Pinker (1994) have gained wide acceptance among scholars; the basic premise of their arguments is that there is a deep neuronal structure in the human brain that corresponds to the syntactic structure of human language. This part of the brain facilitates the acquisition, understanding, and use of language. Furthermore, this cerebral structure actually patterns human modes of thought. That is to say, we think in language.

If it is true that we think in language, then the way that language is constituted will clearly have reciprocal effects on reality. We view reality through the lens of language, and this "constructs"—or to use a more appropriate term, *conditions*—reality. There is a circularity to reality and language, then, whereby language conditions the perception of reality and reality can only be represented through language. Discourse enters into this picture as a way of describing the symbolic representations of thought or material reality. This is done through language and language-like structures: symbols, icons, gestures, and the written or spoken word. Discourse might include texts, speeches, letters, maps, images, or lectures—it is a set of representations of reality that uses language or language-like structures.

Discourse is produced by a particularly situated subject—a subject that is located within a larger social context. The social context is infused with ideology, and the connections between language, discourse, and ideology are deep, complex, and dialectical. Language, as a subset of discourse, structures thought and thereby conditions reality. Like language, discourse also offers a representation of reality, yet it is one that incorporates both language and other symbolic means. It is a set of representations that conforms to certain habitual or learned ways of filtering one's perceptions. This filtering process is best characterized as ideology.

Students of society have been concerned about ideology since the term was first coined during the French Revolution. John Thompson offers a widely accepted definition: "ideology, broadly speaking, is *meaning in the service of power.* Hence the study of ideology requires us to investigate the ways in which meaning is constructed and conveyed by symbolic forms of various kinds, from everyday linguistic utterances to complex images and texts; it requires us to investigate the social contexts within which symbolic forms are employed and deployed" (1990: 7, emphasis added). The literary critic Terry Eagleton, on the other hand, shows how ideology is not so easily pinned down. Eagleton (1991) finds 16 logically separable definitions, and concludes that one is no better than another; rather, a definition must be chosen based on the rhetorical needs of the author and the text.

A sociological definition of ideology would move beyond the epistemological concerns of Marx (where ideology is viewed as illusion or false consciousness, as exemplified in the *camera obscura* metaphor in *The German Ideology*), but also must avoid the relativism of being driven simply by the needs of the analysis. Instead, a sociological conceptualization of ideology should be concerned with social context: the role of ideas and discourse in constructing and maintaining a shared reality, however contested this reality may be. Because of this crucial element of contestation, Thompson's definition of ideology as "meaning in the service of power" is incomplete, for ideology is not just that of the powerful—it is not just about domination. I argue that there is also an ideology of the working class, the subaltern, or the otherwise oppressed. Ideology can be and certainly is used by those in power—to legitimate, naturalize, universalize, exclude, to obscure—in short, to mystify. The philosopher Martin Seliger cuts a line between Thompson's specific definition of ideology-as-domination and Eagleton's relativism. Seliger's sociological definition incorporates both the possibility of domination and contestation: "ideology is a set of ideas by which [people] posit, explain, and justify ends and means of organized social action, and specifically political action, irrespective of whether such action aims to preserve, amend, uproot, or rebuild a given social order" (1976: 14). This definition, by de-linking power from ideology, offers us the ability to speak of an ideology of the oppressed, an ideology of the powerful, a socialist ideology, or a capitalist ideology.

So discourse is a way in which ideology can be expressed but ideology is also an effect of discourse. They are, thus, circular in constituting and reproducing one another. That is to say that ideology (a filtration process) is produced by, but also enables discourse (language or other symbolic meaning systems), while discourse reproduces ideology. This circularity cannot be separated from both the social context—gender, race, and class all play important roles—and the natural context, for the use of nature by society can express systems of domination as well.

NATURE AND SOCIETY: DOMINATION, GENDER, MODERNITY

The use of natural resources is integral to any society. Humans, as with other animals, must harvest materials for use as food, shelter, and warmth. The methods for use, broadly construed as an economy, are one aspect of the relationship that a society has with nature, and an aspect that varies widely across time and space. An economy, however, does not stand alone. A study of political economy will highlight the ways that the state regulates the uses of natural resources, and a cultural study may highlight the religious, mythical, or normative ways that a society regulates its uses of nature. Some method of regulating the economy is imperative, for unrestrained use of natural resources can lead to overuse and crisis.[3]

As a concept, domination often implies force, resistance, and a lack of subtlety. Though it is tempting to characterize any human uses of nature as domination, the expansion of this phrase to cover such a vast territory would render it meaningless. In contrast, 'the use of nature' might imply harvesting natural resources on a sustainable basis. There is, of course, a nebulous area in between domination and use. Damming a river and destroying extensive areas of wildlife habitat, over-hunting of species to the brink of extinction, and the production of radioactive waste which remains poisonous for unknown thousands of years are all candidates for the label domination. But are Native Americans who set fire to the forest to ease the hunting of deer (and to increase the deer's edge habitat) using nature or dominating it? What if the fire rages out of control? The domination of nature must be seen as an ideal type rather than a rigid categorization. Specific examples fit into the rich array of possibility, and can be defined only in relation to one another. In such comparisons, it is important not to romanticize native relationships with nature: though many indigenous groups, such as the Salish in what is now the Pacific Northwest region of the United States, related to nature in a sustainable manner which integrated religion and the economic use of natural resources (White 1980: 14), there are also examples of indigenous humans changing a landscape irreparably, as with the arrival of humans to prehistoric Australia (Flannery 1995: 180).

Sustainability itself is a contested term. Perhaps the most accepted definition of sustainable development is that of the World Commission on Environment and Development's report: "development which meets the needs of the present without compromising the ability of future generations to meet their own needs" (World Commission on Environment and Development 1987). As David Orr points out, by relying on the notion of development, this definition is inherently economistic. Orr argues that the very idea of economic progress may be vitiated by ecological crisis, and offers a (self-labeled) postmodern definition of sustainability: "sustainability requires the transition to a postmodern world that transcends . . . individualism, anthropocentrism, patriarchy, mechanization, economism, consumerism, nationalism, and militarism" (1992: 24). As Schnaiberg (1980) points out, development in the modern capitalist sense is most properly viewed as a treadmill, where once on, getting off is nearly impossible. A move towards sustainability, Schnaiberg argues, would require de-industrialization and other massive economic and political changes.

Robert Brulle (2000: 5), echoing the thoughts of Aldo Leopold, notes that:

> The current social order is the cumulative result of the modern social project. More than 500 years of human activity has been directed towards its creation. The transition to an ecologically sustainable society involves changes at least as great as those that accompanied the transformation of an agricultural to an industrial society. To reverse this cumulative history of destruction of the natural world will certainly take much more time and effort than was previously imagined.

These four ideas of sustainability show how the concept is thoroughly imbricated with modernism—it has primarily been in modern times that irreparable changes have taken place. The rise of modernity, marked by increasing rationalization, urbanization, and industrialization, is largely coextensive with the drastic increase in domination of nature and decreased regard for religious (or mystical) regulation in favor of (generally weak) political regulation.

Max Horkheimer and Theodor Adorno (1944/1993), as with Max Weber before them (1922/1978), linked the destruction of myth and the increase in rationalization with modernity: "The program of the Enlightenment was the disenchantment of the world; the dissolution of myths and the substitution of knowledge for fancy" (1944/1993: 3). Horkheimer and Adorno link this observation to a dialectical critique of Enlightenment thinking in general. To Horkheimer and Adorno, the domination of nature goes hand in hand with the domination of humans by one another: "What men want to learn from nature is how to use it in order to wholly dominate it and other men. That is the only aim. Ruthlessly, in despite of itself [*sic*], the Enlightenment has extinguished any trace of its own self-consciousness. The only kind of thinking that is sufficiently hard to shatter myths is ultimately self-destructive" (1944/1993: 4). Horkheimer and Adorno link the rise of instrumental (means-ends) rationality, via the dialectic of Enlightenment, to the ultimate self-destruction of reason, a situation that seemed imminent in the early 1940s. While this work focuses primarily on human domination of nature, Horkheimer and Adorno's insight that humans also dominate each other in a similar manner is borne out in the empirical cases as well.

This reflection of domination often happens in a gendered manner. Nature, going back to the work of Aristotle, has been characterized as feminine (see, for example Brown 1988: 8), which was often contrasted with the masculine realms of work and politics. As will be shown in the three empirical cases of this work, the discursive construction of the river, as a particular example of a generalized Nature, was associated with the feminine qualities unruliness, passion, and, paradoxically, passivity in the face of human (male) domination. The rivers were constructed as economic resources in terms that echoed the construction of the female role according to the dictates of patriarchal western society. The discursive construction of nature in many ways parallels the construction of *fortuna,* Machiavelli's opposite to *virtu.*

As Hanna Pitkin has argued (1999: 25), *virtu* is manliness in Machiavelli's work: political power is combined with military and sexual conquest. This is contrasted to the effeminate; *fortuna* is the capricious, though generally positive, goddess of fortune or luck (1999: 110). As will be seen in the discourses about nature in the empirical chapters of this work, the fickle and unreliable river (prone to flood and drought) must be mastered by masculine, purposive action. The Colorado River, for example, is labeled as a "menace" by proponents of Boulder Dam. This is meant to convey such unreliability and indeed destruction, as with the Egyptian goddess Isis, whom *fortuna* is often associated with. Pitkin argues that *fortuna* is a woman, and the politics envisioned by Machiavelli is a masculinist domination of the effeminate. Similarly, in the twentieth century American West, male politicians argued for the domination of nature by feminizing the river. The virtuous prince, in other words, would dominate nature through the construction of a dam.

Though Pitkin traces this history back to Machiavelli, and Wendy Brown makes similar arguments as far back as Aristotle, it is with the rise of modernity that the domination of nature becomes a defining characteristic of the relationship of nature and society. As a concept, modernity has its roots at least as far back as the Enlightenment, the Reformation, and the French Revolution. As it is often used today, the term describes an ideology that enshrines instrumental reason as a guiding force in all human affairs. While Weber saw modernity as being rooted in the twin processes of rationalization and secularization, the roots of this concept owe to the Enlightenment thinkers who founded political liberalism; to Luther and the Reformation who individualized the relationship to God, the Calling, and salvation; and to the French Revolution, which brought rationalization to many other areas of society.

The Reformation must be taken as one of the important moments on the road to modernity. As Weber (1905/1958) has persuasively argued, Luther's concept of the calling and the individual relationship to God and salvation was important to the development of capitalism and instrumental rationality in the West. Habermas (1987: 17) further points out that the key to Luther's faith was that it was reflective, that is, the individual subject relied on his own insight rather than just that of the religious authority. Similarly, the individual was at the core of both rationalist and empiricist early modern philosophies, such as those of Descartes, Liebniz, Locke, and Hume. To brush with extremely broad strokes, many of the Enlightenment thinkers saw the individual as a foundation for structuring a theory of being. Rene Descartes' famous *cogito ergo sum* argument posited the individual doubter as the basis for an ontological solution to skepticism. G. W. F. von Liebniz similarly argued in his *Monadology,* that the rational individual was the center of his universe. The political philosophers of the time (e.g. John Locke, Thomas Hobbes, and David Hume) derived political and moral rights from the individual and his/her relation to the state of nature. Social contract theory, such as that of Rousseau, was the ingenious way that some thinkers could ethically subordinate the individual's rights for the good of society. The French Revolution, based in part in such political philosophies, attempted to institutionalize such a balancing act (Descartes 1641/1974; Hobbes 1660/1982; Hume 1748/1974, 1751/1988; Liebniz 1714/1974; Locke 1690/1974; Rousseau 1761/1988).

The French Revolution, according to many theorists, is truly the gateway to modernity. The philosophers and political activists of the Revolution attempted to rationalize all aspects of society: time, weight, measure, government, and even culture. Though the Revolution was—as is always in history—a complex process of change, it helped to bring together the ideas of the individual as the rights-bearing citizen and the idea of the rationalization of all aspects of society.

Lynn Hunt's (1984) cultural history shows that the Revolution was forever changing. This revolution-as-text was its own hermeneutic, Hunt argues, and the only unity that arose from it was the idea of "newness"—the idea that what was going on was a fundamental change, that an entirely new state was being constructed. Out of this arose the new revolutionary calendar, the metric system, and new political symbols, including the liberty tree, particular styles of clothing and adornment used to indicate a political stance (the *sans culottes*), and the female figure of Liberty. Though continuities with the Old Regime were apparent, as Sewell (1980) has shown, the French revolutionaries were attempting to follow Rousseau in a radical direction by truly adhering to the general will in fashioning an entirely new—and modern—social contract.

Modernity since the French Revolution has become inextricable from industrial technology and the development of capitalism. Modernity, according to Weber, involves the rationalization and demystification of the world (1905/1958: 26; 1946: 293) especially on the level of government and economy. But, according to Habermas, it also occurs on the level of the individual subject; modernity involves the "development of modern societies from the viewpoint of rationalization," and the "patterns of socialization that are oriented to the formation of abstract ego-identities and for the individuation of the growing child" (Habermas 1987: 2).

The concept of modernity thus includes many of the watchwords of sociology: rationalization (and concomitant demystification), industrial and political revolution, socialization. But all of these ideas and processes are clouded and crosscut with contingency, as history is a process of both continuity and change. Modernism came to its fullest expression in the late nineteenth century, but this is precisely the time when Jackson Lears (1981) takes note of antimodernist tendencies in the west.

Lears characterizes antimodernism as a search for medieval or "oriental" alternatives. It is "a longing for a regeneration at once physical, moral, and spiritual" (1981: xii). As such, antimodernism is partially backwards looking. Yet Lears also identifies Theodore Roosevelt's "cult of the strenuous life" (1981: xii) as a manifestation of antimodernism as well. How can the famously progressive president be both progressive and antimodern? The answer, for Lears, is that the "antimodern sentiment was unstable, ambivalent" (1981: xii). It represents the yearnings for that which is lost in the march of rationalization and demystification. Antimodernism is, in part, the carry-over from times past, the continuity in history's dialectic of change. Antimodernism, then, can take on many faces and constitute, in a paradoxical manner, an easing of the transition to modernity.

The role of the state in the uneasy transition to the modern age will be addressed in the next section and the empirical chapters beyond it. Much of this transition to modernity was facilitated by both the state and civil society, in part based on the construction of a nature-society relationship that enabled domination of nature.

POLITICAL SOCIOLOGY: STATE POWER, NATURE, AND TERRITORIALIZATION

The treadmill of production has been a powerful concept for analyzing state and economic action from an environmental and political-economic perspective (Schnaiberg, 1980; Schnaiberg and Gould, 2000). In this view, growth is predicated on resource use: availability of land, water, and power are crucial, as are the needs for labor and capital. The concept of the treadmill of production has been used primarily to analyze established economies—

cities looking to expand, states in need of development, economic managers seeking higher rates of growth. But the treadmill of production may also be a useful concept for analyzing the manner in which young states engage in development and expansion, on frontiers as well as in cities. The forces that effect state building processes include class conflict, internal and geopolitical crises, increases in bureaucratic or technical control, and the need for economic growth. How the state understands the landscape, interprets the availability of resources, and reacts to perceived barriers or benefits in the natural world also helps to define how growth will proceed.

Allan Schnaiberg gives the examples of the Progressive Era and the New Deal as moments where the state has acted in the interests of capital or labor to make corrections or to provide stability. Mostly he focuses on the last 50 years, in which the economic growth models have formed the basis of the ideology behind the treadmill of production (Schnaiberg 1980: 217). Schnaiberg focuses on the economic sphere to important effect: "the basic social force driving the treadmill is the inherent nature of competition and concentration of capital" (1980: 230). But this logic was also clear in examples of state building in the Progressive Era and the New Deal in the American west.[4] It is such state payouts that this paper will focus on, as they provide an onramp to the treadmill of production for developing states and regions.

The treadmill of production is something that societies may mount very early—this is to say, the treadmill is not just a phenomenon of a modern, or urban society. Young states can be subject to the legitimation and fiscal crises that O'Connor (1973) and Habermas (1989: 289–91) speak of, and states must thus get onto the treadmill in order to solve them. Schnaiberg recognizes this and asserts historical and cultural variability. Once on the treadmill, advanced capitalist economies (and in existing socialist economies) the treadmill of production has a unanimous coalition of groups behind it that are more or less unassailable.

Schnaiberg shows that the solution to fiscal crisis lays in the mutual interests that the state, capital, and labor in growth. Through these mutual interests the entities cooperate in mounting and speeding up the treadmill of production. Of course these entities also have various conflicts over resources, taxation, and policies like a minimum wage, entitlement programs, and pollution controls. As Marx pointed out in *Capital* in the sections on cooperation, there is a dialectical dance between labor and capital (and now one would add the state) in the process of working out how to optimize the political-economic system (1867/1967: 333).

An interesting expansion upon Marx's notions of cooperation and use value versus exchange value can be seen in Logan and Molotch's work on the urban growth machine. Logan and Molotch identify the urban growth machine as "nowhere more clearly documented than in eighteenth- and nineteenth-century American cities" (1987: 52). Especially important during these eras were federally-funded or -subsidized transportation infrastructure improvements like canals and railroads. In the American West, just outside of the cities, the hinterlands that would funnel the timber, livestock, and agricultural produce into the city markets were equally dependant upon the federal government to control natural resources so as to enable economic and social development. As then Commissioner of Reclamation Arthur Powell Davis said in 1924, society needed to convert rivers from being "a natural menace to a natural resource" (Davis 1924).

So linking growth to water supply and other environmental factors has long been seen as important. All states have to deal with some form of scarcity, whether it is land, water, transportation, labor, or capital. To use Schnaiberg's example of the Progressive Era in the United States (1980: 217), resource development by the state in the face of scarcity was seen as the solution to capital's inability to develop an adequate foundation for growth. The concept of the treadmill of production can be used to examine this state-building process. As the state acts to overcome initial environmental barriers to growth (especially those that capital has failed to deal with), the state engages in various roles in helping or hindering the development process. The state's interests in avoiding both fiscal and political crises were served by the development of the western frontier in the early part of the twentieth century: by developing resources that capital could not, the state encouraged settlement of the region while providing the water and flood control needed for agriculture and an expanding population, as well as energy for industry, and jobs for an underemployed labor pool.

Adapting the treadmill of production to better understand the state-building process as a machine for growth involves examining the complex role of the state in modern sociology. In recent years, the state has become a focal point for many academic discussions, especially those within political sociology.[5] The correction of overly society-centered or class-centered approaches, most poignantly articulated by Theda Skocpol, has had important implications for theory and empirical research (Burawoy and Skocpol 1982; Finegold and Skocpol 1995; Domhoff 1990; 1996; Skocpol 1979; 1985; 1992; Gilbert and Howe 1991). Analysis of the state, however, is questioned by Foucault, who suggests that such studies have led the concept to become hypostatized, and that it is better to investigate the way that the state has become governmentalized. Governmentality, to Foucault, involves the institutions, procedures, analyses, and tactics that allow the exercise of state power. The eventual pre-eminence of this form of power has led to specific governmental apparatuses and a complex of knowledges (Foucault 1991: 102).

The study of the state, Foucault contends, often descends into either a description of the monster of bureaucracy or the simple reduction of the state into a series of functions. Foucault argues that a more interesting question is to examine the way that sovereign society was replaced with disciplinary society, which has now been replaced with a governmental society. Foucault argues that "the state, no more probably today than at any other time in its history, does not have this unity, this individuality, this rigorous functionality, nor, to speak frankly, this importance; maybe, after all, the state is no more than a composite reality and a mythicized abstraction, whose importance is a lot more limited than many of us think" (Foucault 1991: 103). Perhaps Foucault is correct in suggesting that scholars are overly fascinated by the state.

But the effects of the state are real. The state does have the power to implement its authority over a land and a people—to enforce upon society a more disciplinary or a more surveyed existence. While Foucault is largely correct in pointing out the broad changes from a disciplinary society to one characterized by governmentality, this level of abstraction does not explain the causes of certain state actions—the state, after all, is not monolithic. States and their various constitutive agencies and actors behave differently in different times and places. Given an understanding of the reality of the state's power, I will heed Foucault's warnings and avoid a simple reduction of the state to a series of functions. Instead, I want to explore the ways that this fragmented yet composite reality that we call the state has been built up in the American West. In particular, state building can be understood as a combination of the ways that parts of the state view the landscape and the roles that the state takes in relating to nature, and this combination has important implications for how nature and society interact.

State building[6] is discussed in a rich literature within political science and political sociology. Many scholars of the American state have attempted to reduce the complexity of the state and its context to a series of binaries or a simple main cause: American exceptionalism from Europe, the response to geopolitics, or the reaction of the professional class to urbanism and political corruption. Stephen Skowronek (1982), however, has underlined the complexity of the state building process that is missed by such conceptions, and has posited a tripartite schema where the three elements interact in intricate ways. American state building, Skowronek argues, involves 1) officials' responses to external stimuli, which characteristically take the form of domestic or international crises, 2) class conflict, or 3) evolving complexity due to the rise of industrialism after Reconstruction ended in 1877 (Skowronek 1982: 10–11). Rather than reducing the state to an effect of class relations, or the result of geopolitical forces, Skowronek allows such forces to interrelate in shaping the American state. This is an improvement upon more simplistic understandings of state building because it allows for the contingencies apparent in the historical record—especially those series of events that are difficult to theorize. But Skowronek's analysis relies heavily upon the institutions of the eastern portion of the American state. State building in the American West, on the other hand, is less a response to class conflict or geopolitics (though these play important roles) than it is an attempt to establish social control and stable political governance in the face of a complex set of environmental barriers—in other words the state mounted a treadmill of resource exploitation for growth. The building up of the American state in the West involved the creation of a treadmill of production where the state supplied the infrastructural basis for economic growth and social stability. In this chapter, I will focus on the state's construction of large-scale water systems as an empirical instance of this process, but examples abound: the establishment of a standing army in the U.S. to control the indigenous people and eject other imperial occupiers (France, Russia, England, Spain, and then Mexico), the land grants of the Homestead Act, the encouragement of railroad expansion through similar subsidies and land grants.

Traditional understandings of state building, which Skowronek argues against, focus on the growth of empires in early modern Europe. As a nation's sphere of influence expanded, it would begin to chafe against other nations. Various understandings of this process have been propounded (see for example Anderson 1974; Ertman 1997; Tilly 1990; Wallerstein 1974). This geopolitical focus is retained in Skowronek's work, but he does not discuss the nineteenth- and twentieth-century American West. The West had long been an incubator for bureaucracy and federal involvement in local affairs, with a strong military presence being perhaps the clearest example (Abbott 1994; White 1991). In the Progressive Era, this strong foundation yielded a case of state building that does not easily fit existing theories of state building, though the process was undertaken using the standard techniques of an expansion of administrative capacity and the establishment of an infrastructure for continued economic development. As the empirical work of this book will show, in the twentieth-century American West, the state continued to build a political and economic infrastructure in response to the chaotic development and fragmentation that had resulted from westward expansion.

The process of state building in the West had a complex set of environmental conditions to overcome, and the way that both local actors and the state viewed nature in the West had important environmental implications. Not only was there a much wider expanse of land to deal with in the West than had existed in the eastern regions—what Valerie Kuletz (1998) calls "vastness"—but the terrain was also diverse in geographic character and climate. There were problems of water scarcity, rugged terrain, exposure to the elements, a lack of timber for housing and fuel; in short, all of the problems of the frontier were difficulties not only for individual families, but also for the state. Perhaps primary among the state's roles in the eighteenth- and nineteenth-century West was the protection of settlers from Indians. Given the geographic problems of the territory, the army had to become an occupying force whose ultimate goal was the complete removal of the Indians (White 1983). All of these barriers and problems also had to be overcome in the East but by the end of the eighteenth century it was largely a completed process. In the West the process continued well into the nineteenth century; indeed, some argue that it continues today, though in altered form.

This difference in the process of state building was largely an outgrowth of the different responses that the state had to variations in geography. These responses are based on a particular ideology of the domination of nature that is similar to the ideology of the eastern regions of the U.S. But it is an ideology that is driven to extremes, both by the extremes of the landscape and the extremes of the state's responses to that landscape. States have a peculiar power to enforce authority over not only a people, but a territory. This state power is garnered through a monopoly on the means of coercion (Weber 1922/1978), as well as a process of information gathering that James Scott (1998) calls "state simplifications." In a twist on Frederick Jackson Turner's (1920/1996) hypothesis, I argue that the culture of the frontier and the important changes in class structure that occurred through the twentieth century are important parts of how state building proceeded in the American West. Without the reinforcing processes of state simplifications and other cultural constructions of the landscape that occurred in the public sphere, state building in the American West would have been much more circumscribed.

State Building Through State Simplifications

In *Seeing like a State,* James Scott gives some examples of the interconnections between state power and territorial control that are a useful model for understanding state building. Scott asserts that the early modern states of Europe redefined their role in the face of alarming and complex changes in society. Given the logic of bureaucracy and the ascendance of rationality, the state's preferred methods of re-organization were increases in knowledge and control (Scott 1998: 13).

The administrative ordering of nature and society was achieved though the mapping of territory, the insistence on surnames for taxation purposes, the imposition of standard units of weight and measure, the enclosing of territorial boundaries, and the enumeration of the state's holdings. In one example, Scott is specifically interested in the ways that the German State regarded its forests; he discovers strong tendencies toward rationalization and technical knowledge. The state narrowed its vision of the forests to a viewpoint that saw only commercial uses and attempted to bring the forests under control to intensively develop those uses. In phrasing that sharply evokes Marx's ideas of how exchange-value supplants use-value in the transition to capitalism, Scott shows how the state foresters came to see the multiple uses of the forest by the peasants as inappropriate. Scott's Weberian twist is that the state (rather than a capitalist group) is the "seer," and the state has the power to impose its own logic on the land, thereby changing the societal understandings and uses of nature. In this state logic, trees become lumber, animals become game or pests, plants become weeds, and rivers become pipelines.

German foresters rationalized and standardized the forest to make it more productive. Over time, the old-growth forest was intensively managed to produce the *normalbaum,* or standardized forest, which is a mono-crop of trees in straight, orderly rows. They are of the same age and are scientifically measured, categorized, and managed for a maximized and non-fluctuating yield. This management technique not only increased economic returns and created a park-like forest; it also simplified labor and training, in effect easing the process of disciplining the foresters.

The result of this endeavor would be immediately clear to a modern ecologist. The built-in problems with these management techniques are common to all forms of mono-cropping: single species in large numbers provide ideal hosts for insects and disease, thus increasing the likelihood of catastrophic outbreaks. The German forests eventually lost much of their capacity for sustained yield, but the damage had been done. The state had altered the natural landscape, imposed a strict, rational order upon its territory, and in the process had created a trans-national phenomenon. Gifford Pinchot, the future head of the U.S. Forest Service, learned from these innovative Germans and brought their techniques home to Progressive-Era America.

Pinchot wanted to manage America's vast forest resources using scientific techniques for sustainable yields. He argued the utilitarian idea that resources should be managed for the best use for the longest time:

> The purpose of Forestry, then, is to make the forest produce the largest possible amount of whatever crop or service will be most useful, and keep on producing it for generation after generation of men and trees (quoted in Worster 1994: 267).

Perhaps most important, these techniques did not remain exclusively within the realm of forest management; the ideas that motivated these techniques resonated with other branches of the federal government during the Progressive Era—especially the U.S. Department of Interior's Bureau of Reclamation. In 1902 the Bureau of Reclamation (called the Reclamation Service until 1926) was charged with the task of reclaiming arid lands in the West through the construction of dams, canals, and other large-scale irrigation projects. To do this efficiently the Bureau needed to see the rugged terrain of the West through the eyes of an engineer; it needed to see the West as the German foresters saw their *normalbaum*—a rational and homogenous landscape.

Scott uses the forestry example as a point of entry for talking about "state simplifications" (Scott 1998: 80). The state "sees" a dynamic world and imposes its logic to create static images on the cadastral map. These state simplifications collectivized, standardized, and measured only subjects of official interest, such as taxes, conscription, or social control. Though some amount of state simplification may be required to run a nation (such as taxation or conscription), Scott goes on to describe how some state simplifications can be over-extended: just as the German state misunderstood forest ecology with the *normalbaum,* other states have erred in well-meant efforts to improve the social condition. State simplifications lead to failure when a weak civil society cannot oppose an authoritarian state, especially during the reign of what Scott calls the ideology of high-modernism.

Scott (1998: 89) defines high modernism as

> a strong version of the beliefs in scientific and technical progress that were associated with industrialization in Western Europe and in North America from roughly 1830 until World War I. At its center was a supreme self-confidence about continued linear progress, the development of scientific and technical knowledge, the expansion of production, the rational design of social order, the growing satisfaction of human needs, and, not least, an increasing control over nature (including human nature) commensurate with scientific understanding of natural laws.

As will be shown explicitly in the following empirical chapters, the Bureau of Reclamation exhibited many of these goals and beliefs. The Bureau operated by simplifying the dynamic arid landscape of the American

West to impose an infrastructure that would build up state institutions, increase social and political control, and provide the basis for economic and population growth. Although this is by no means the only way that the Bureau of Reclamation could have viewed the landscape of the West, this ideology was over-determined due to characteristics of Progressive-Era American government, as will be explored in the next chapter.

Scott's book is devoted to how plans for social engineering on a grand scale have failed. Scott identifies four factors that explain this failure: the administrative ordering of nature and society (such as the *normalbaum*), the ideology of high modernism, an authoritarian state, and an anemic civil society (Scott 1998: 4–5). In the presence of the ideology of high modernism and a large and ambitious state, a weakened civil society cannot oppose state domination. The cases examined in this work, however, show that an active and vibrant political culture and public sphere in the American West conforms more to Frederick Jackson Turner's hypothesis (1920/1996) that the frontier was the fount of democracy than to Scott's idea of high modernism. A "prostrate civil society," as Scott calls the authoritarian high modernist public sphere, is not an accurate description of the American West in the early twentieth century.[7] While some version of the ideology of high modernism is, I argue, hegemonic for at least the first half of the twentieth century, its hegemony involves an "impure" example—there are several different inflections of high modernism apparent in the early twentieth century American West. As will be demonstrated in the subsequent empirical chapters, different versions of the ideology of high modernism contributed to varying outcomes in the domination of nature.

Scott's several other factors that lead to failure—a strong state, a weak civil society, and administrative ordering of nature—were also not all present in the American West. A strong civil society, this work shows, called upon the state to build an infrastructure, rather than a state imposing it through authoritarian measures. In fact, by the 1960s, a resistant civil society stopped the state's plans for further development. The explanation for the rise of an effective opposition is found, in part, in the rise of the middle class in the mid-twentieth century United States.

Scott's inattention to temporal and geographic variation while focusing only on negative cases (Coronil 2001) has had the result of leaving a class analysis out of his work. Scott's focus on the actions of the state in his several cases serves his analysis well—the radically different class structures of his different examples would deepen the complexity of an already complex thesis without offering a concomitant increase in explanatory value. In the cases presented here, however, class analysis is important for understanding the support or resistance to state actions. Many of Scott's ideas are applicable to the American West, especially the administrative ordering of nature by a powerful state that simplifies the landscape and applies the ideas of technology and science. But the ideology of high modernism is not monolithic, and a strong state was matched at times with a strong, middle-class based civil society. Thus, the case of the American West does not lead to an utter failure on the order of Brasilia, Tanzania, or Soviet collectivization.

One of the reasons why Scott identifies this state-simplification process as leading to failure is that while the state may view this state-building process as a simple one, state building actually reflects the dynamic interactions of nature and society. State simplifications have had far-reaching consequences for the environment, but at the same moment nature has both enabled and constrained the ways that society has been shaped by state-building processes: Nature and society exist in a dialectical tension where each mutually constitutes the other. Scott examines several historical cases that challenge the traditional view of state building, but his cases are at times difficult to generalize from because of the complexity of geopolitical dynamics and the contingency of events. The variations evident in the case of the early twentieth-century American West offer a further basis for questioning the generalizability of Scott's ideas.

Scott's high modernism is defined only by the cases that lead to failure (Coronil 2001), and other examples show how ideology varies over space and time. The extremity of Scott's cases help to support his arguments, but as the ideology of high modernism spread through the political culture of the U.S. during the Progressive Era and the New Deal, characteristics of the ideology were tailored to regions and specific programs or agencies. As the empirical chapters of this work will show, high modernism in the West was uneven and not always totalizing. As Lears (1981) argued for modernism in general, high modernism is exhibited by different social groups and is cross-cut by antimodern sentiments. There were also social groups who were avowedly anti-high modernist. Several social groups in the American West were adamantly opposed to the state's involvement in building an infrastructure. The conflict between these groups is the main subject of the next chapters.

The process of state building in the American West certainly was characterized by the administrative ordering of nature and society, and this was accomplished by a strong state that acted, at times, with fierce authority. The ideology of high modernism is also exhibited, though in impure and varying forms, but there was not an anemic civil society. To the contrary, the civil society of the West was a vibrant force that invited the state to help build an infrastructure to support the economy and control the territory. Such complicity in the ordering of nature and the process of state building on the part of civil society shows the interaction of the state and the public sphere in the domination of nature.

In order for the state to build an infrastructure in the American West, certain justifications had to be made. To build a dam, the state not only had to "see" the landscape and map the territory in certain ways, it also had to legitimate its actions. Such legitimations and justifications largely occur through cultural processes, with distinctions by class. In the case of the American West, natural formations such as rivers and canyons had to be discursively constructed as economic resources to justify the creation of dams and reservoirs, and state actors and local elite groups constructed nature in very different ways than Native American groups or opposition groups. In general in the early cases, the cultural practices of local groups helped to discursively construct nature in general—and rivers in particular—as natural economic resources rather than as entities that carried value in and of themselves. In later cases, effective opposition from middle-class groups helped construct nature as having inherent value.

But the control of territory does not simply exist on a material plane: cultural tools such as maps using the imaginary grid of longitude and latitude allowed the state to "see" the landscape in a certain way, and this allowed for the rational understanding and efficient use of resources. Cultural processes exert pressure on how the state institutes its control. In the American West, the culture, or myth, of the frontier played an important role in conditioning the state's responses to issues of social control. Similarly, Chandra Mukerji's study of France under Louis XIV illuminates the cultural aspects of the ways that states "see" (Mukerji 1997).

Louis XIV's massive project of state building was an example of France's territorial ambitions. The goal was to mark off a territorial boundary: France carved up and transformed nature at the edges of its territory. The royal engineer built earthworks and river diversions to enhance the border when natural features weren't enough to make delineations clear. These boundaries were required in order to give form to France, a firm boundary for the newly forming modern state. These techniques were "showcased" at Versailles: gardens, water-works, terracing, and statues proclaiming French taste. Nationalistic and boundary-forming cultural objects were in dynamic tension with the exhibition of exotic plants that showed the reach of France's trade networks. The trade networks showed the power and influence of France—but also her dependency on others in a globalizing economy.

Mukerji argues that the gardens of Versailles served a cultural and ideological purpose for France under Louis XIV. The gardens served as a mirror of the King's power: they reflected the control he had over the land and the people of France. In the same way that the gardens of Versailles represent the territorial ambitions of Louis XIV, the dams of the twentieth-century American West represent the ideology of high modernism: bureaucratic planning, state administration, and the social control of nature. Material culture, such as gardens, waterways, and dams, can show the importance of myth and symbol in the display of state power. Culture, however, also operates in direct ways in creating a discursive foundation upon which the state can build.

State Building in the American West

To overcome the natural formations in the American West the Bureau of Reclamation invented a series of engineering marvels, from Boulder Dam to the All American Canal. Nature has been slowly undoing this work over the last century through the build-up of silt behind dams due to erosion. According to geologists and the Bureau of Reclamation itself, the dams of the West will likely outlast their reservoirs. As silt from the slow erosion of the mountains of the West builds up behind the dams, eventually the dams will hold back reservoirs of silt (Bureau of Reclamation 1961; Reisner 1986). The rivers will simply run over the tops of the dams in spectacular waterfalls. Such a dynamic between nature and society—where nature undoes human work—is often not readily apparent. This book endeavors to show how the unintended consequences of relations between nature and society can effect the relations between humans. The discursive construction of nature—in this case a river—as, simultaneously, a tyrant, a menace, and a wasted resource, resulted in the domination of nature in the form of a dam. The limited lifetime of these dams will have unintended consequences in the future: The state's building of an infrastructure that is not sustainable will come back to haunt the citizens of the West. The simplified view of the land- and water-scape of the West has been successful in the short term at stimulating economic growth, providing the basis for an expanding population, and has secured the territory of the West for control by the state. The state's simplifications in the American West (in the form of dams, water ways, abstract grid-based maps, and property regimes) provided large subsidies for capital (and for a few small family farmers). The state supported capital in addition to its own direct interests, and operated as an engine for economic and population growth. But the limits on this development will be reached. How society deals with these limits, though an open question, will likely involve continued domination of some social groups by more powerful ones.

Donald Worster, a present-day environmental historian, has succeeded in laying out a model for understanding the role of infrastructural aspects in conditioning the way that power is distributed in a society. Worster examines in detail how power, economics, infrastructure, and ideology are interwoven. He works explicitly from the ideas of rationality and domination that were developed by Max Horkheimer and Theodore Adorno. Worster uses these concepts to explain the modern "Capitalist State Mode" of domination. Here there is "no single despotic ruler who personifies human control over the environment. In the new mode, power becomes faceless and impersonal" (1985:52).

Since Worster works out of the theoretical framework of the Frankfurt School, it is not surprising that he uses a theory of power that matches critical theory very closely. Building on Karl Wittfogel's (1957) *Oriental Despotism,* Worster enlarges the scope of the Frankfurt School to include infrastructural elements, such as the large hydraulic projects of the American West. Control over these projects amounts to control over the majority of the economy and society. This control is shared between the bureaucratic technical experts and the private agricultural sector that benefits most from the water. As Worster (1985: 261) explains it, as the modern capitalist state mode of domination

> came to maturity [in the American West], its structure was revealed to be one of a small power elite reigning over a large, anonymous, dependent population. That elite had both a public and a private face, the double-sided face of the modern capitalist state.

Worster's argument, however, goes too far. This mode of domination is constrained by democratically controlled law in the western United States (even if this political system is distorted by power and influence), a constraint that Worster appears to dismiss in his text. In his conclusion, however, Worster returns to popular democracy, and elevates it as a panacea for the problems of the modern capitalist mode of domination (Worster 1985: 332). Certainly we may argue that irrigation and large-scale water systems have an important effect on economy and society in the arid American West. But it is too drastic to argue that the American West approaches Wittfogel's term "oriental despotism."

So how did the state alter the environment in the American West, and how did nature similarly help to shape state-society relations? Such a question must combine the ideas of political sociologists, environmental sociologists, and environmental historians to arrive at a mutual informing of theory and history. This integration is rare, but many environmental sociologists, historians, and geographers have called for such a new paradigm within social theory.

ENVIRONMENTAL SOCIOLOGY, ENVIRONMENTAL HISTORY, GEOGRAPHY: THE DIALECTIC OF NATURE AND SOCIETY

Much of the sociology of environment has developed in response to industrial pollution (especially air and water pollution), natural resource extraction and their dependent communities (the effects of mining and logging), and the impact of growing numbers of humans upon the globe in general (carrying capacity), and some regions in particular (the rich North blaming the impoverished South). Some environmental sociologists have examined agriculture, but have mostly focused on pesticide and herbicide use, the effects of this upon land fertility, the effects of mono-cropping on eco-systems, and the health of the farm laborer.

What all of these sociological conceptions of nature and society have in common is their separation of society from nature, though many times this is done for analytical ease while recognizing at some base level the complexity of their interrelationship. What I would like to argue is that this separation is untenable. If one looks at nature and society on a historical level it becomes clear that there is so much mutual interaction as to vitiate any easy distinction. As Raymond Williams has noted, "nature" has an extraordinary amount of human history in it (1980).

In "Beyond the Nature/Society Divide," Freudenburg et al. (1995) argue that a fruitful approach to understanding the complex web of causality in environmental sociology is to build upon insights such as Williams' above. Society and nature are not so easily separated, the authors argue, and instead should be viewed as "conjointly constituted." Freudenburg et al. trace the history of a Midwestern mountain (Iron Mountain in Michigan's Upper Peninsula) to exemplify the different social perceptions of the mountain and its uses, while showing the essentially unchanging qualities of the mountain itself during a period of human history. I wish to build from their work by showing the ways that the landscape and human perceptions of it are mutually constituted, and how this mutual constitution can, in a cumulative and circular manner, lead to further significant alterations of nature.

The authors demonstrate how social perceptions of the natural world influence the ways that humans interact with nature. Just as different cultures have viewed water in varying ways, different cultures have also valued mountains differently, and this has led to varying uses of the mountain. Indigenous people did not use the mountain in the same ways that French trappers or American lumber barons did. Indeed Indians could not have done so, for the social relations in which they lived were significantly different from the imperialistic European social systems. It was the set of social relations, the technology, and the economy of the Anglos that created the possibility for viewing the mountain as an iron ore-bearing resource.

Soon after Iron Mountain came to be seen as an economic resource, the ore was mined and the landscape was altered. This is a point that the authors do not give enough attention to. Social perceptions of the landscape can lead to intrusive or benign action; they can be ecologically sustainable or not.[8] While it is important not to romanticize the understandings of the landscape that indigenous people held, theirs was often a more sustainable relationship. The American way of seeing the landscape (as compared to native or French colonialist ways of seeing), when attached to the cultural and economic practices and massive military power, triggered significant changes in the landscape. By encouraging individuals to mine the iron ore, the American social system (economy, technology, property regimes) was responsible for altering the physical landscape in ways that centuries of erosion or other non-human physical processes did not. Similar processes are at work in the nineteenth-century American West. As this work will explore, the Anglo way of "seeing" the landscape of the West was a prerequisite for the state building projects that would further change the landscape from a realm of wild rivers to a set of tamed, slack-water reservoirs.

Freudenburg, Frickel, and Gramling's attempt to move past the nature/society divide shows what is missing from much of the discussion of society and environment: the specific ways that nature provides resources for humans, while humans simultaneously produce and reproduce understandings of nature. O'Connor (1998:46) points out that:

> Culture and nature meet and combine in socially organized social labor. Cultural ecology and ecological culture are expressed in the social relations of material production, distribution, exchange and consumption. The question arises, What is the dialectic of nature and culture in material life generally?

The imposition of human-built structures upon the environment, the selective breeding of plants and animals, the redirection of waterways for irrigation, even climactic changes due to erosion, deforestation, regeneration, or clearing through fire all are significant examples of the dialectic between humans and nature. The sociology of environment needs to move away from a separation of society and nature, even if this separation is only for analytical purposes. In the final portion of this chapter, I combine the ideas of environmental and political sociology with those of environmental history and geography.

Dialectics of Nature: History, Geography, Sociology

The insight that humans are a part of nature is not new, and it can be construed as not much more than a banal truism. We are not exempt from our environment, but a cursory reading of most sociology journals would lead one to believe just that. As William Catton and Riley Dunlap (1978) have pointed out, we live in a "Human Exemptionalist Paradigm." But if most sociologists would agree with the truism that humans are not exempt, then why employ the separation that is a starting point for so many analyses? Even a non-ideologically motivated attempt to separate society and nature, perhaps justified by the need to simplify an analysis, will fly in the face of the historical evidence suggested by environmental historians such as William Cronon, Donald Worster, or Richard White.[9]

The idea of nature and society existing in a dialectical relationship goes back at least to the insights of Karl Marx, who in *The German Ideology* (1932/1972:174) writes,

> He [Feuerbach] does not see how the sensuous world around him is not a thing given direct from all eternity, remaining ever the same, but the product of industry and of the state of society; and, indeed, in the sense that it is an historical product, the result of the activity of a whole succession of generations, each standing on the shoulders of the preceding one, developing its industry and its intercourse, modifying its social system according to the changed needs. Even the objects of the simplest sensuous certainty are only given him through social development, industry, and commercial intercourse.

Marx is arguing that one must probe beyond the apparent material reality that Feuerbach examines, and in the lines that follow the above passage, we get an interesting intimation regarding nature and history. "The cherry tree," says Marx (1932/1972: 174),

> like almost all fruit trees, was, as is well known, only a few centuries ago transplanted by commerce into our zone, and therefore only by this action of a definite society in a definite age it has become a 'sensuous certainty' for Feuerbach.

Marx has anticipated many of the arguments of the present-day environmental historians regarding "second nature" (Cronon 1991). What is taken to be a concrete natural feature is in fact mediated by human activity.

This mediation occurs through the transformation of nature by the hand of humans, primarily through labor within a capitalist system. Geographer Neil Smith's (1984) concept of the production of nature provides a starting point to understand how the transformation of first nature into second nature is the unintended consequence of capital's search for new realms from which to extract surplus value. This implies that those areas from which more profit can be squeezed will be utilized first. The production of nature must follow the other laws of the capitalist mode of production, and so the development of nature occurs via the profitability that is expected. In the American West nature has been produced in part due to these pressures from private capital, but also in part due to the exigencies of the state. As Robert David Sack has argued in his meditations on territoriality, the links between the nation state and capitalism have a tremendous impact on the way that territory is understood (Sack 1987).

Harvey (1989) extends the argument regarding the production of nature to the production (and compression) of time and space: these social constructions are, as Marx argued, annihilated due to political-economic forces. Harvey takes a materialist perspective by arguing that time and space have no separate existence from materiality, thus, we cannot understand history or geography outside of a political-economic context. While the state gains political, ideological, and military power through control of space, capitalists gain economic power by controlling labor time for workers.

Just as the state seeks to annihilate space through the construction of roads, airports, and telecommunications infrastructure, capitalists seek to increase the turnover time of capital to extract more surplus value. Harvey uses these ideas to construct an argument regarding the character of the present moment and its continuities with the past. But his argument, based so strongly from Marx's ideas in *Capital,* are just as applicable, perhaps on a different scale, to the entire period of modernity. Private capital has certain affinities with the state, one of which being an interest in the compression of space and time.

In the American West, the affinity between the state and private capital yielded an infrastructure that helped to compress the vastness of the territory. The variability of the landscape, however, has hindered the total compression of time/space, as natural conditions vary over space and through history. This material reality is inescapable: Time and space exist only as referents to actual physical things, such as the time it takes a worker to make a widget or the distance and transportation time from farm to market. Some of these physical aspects of space and time put barriers on the compression by capitalism, such as the aridity of the American West barring the use of the fertile fields and long growing season. In the face of this complex landscape that resists easy compression, the state was forced to act where private capital could not. To achieve the goals of state building—economic growth, population growth, and social control—the state had to simplify this dynamic landscape, in the manner that Scott (1998) has noted.

The state, then, has its own set of interests in compressing space and time, in bounding a territory, and especially in promoting economic development. In cases of long-term investment such as that required to overcome large ecological barriers, the state often acted where capital cannot. Such actions as building dams to enable further capitalist accumulation represent a "spatial fix" for recurrent capitalist crises. David Harvey (2000: 23–30), building on Marx (1867/1967), argues that colonization or imperialism offers a way out of crisis through the appropriation of new spaces of accumulation. As we will see in the empirical chapters of this work, just such a developmental process occurred within the western region of the U.S. in the early part of the twentieth century. Though labeling of this process as internal colonization, imperialism in one country, or some other phrase is difficult due to the thorny issues of who might be colonizing whom, the actors at the time called their projects the "completion of empire." We may thus understand the development projects—at least from the perspective of the historical actors and from the contemporary analysis of the role that such developments play for capitalism writ large—as constituting a spatial fix for capital.

Though greatly political-economic in character, the spatial fix process also has cultural elements. Cultural constructions of nature, society, time, and space are both based in and translated back into materiality. David Harvey's *Justice, Nature, and the Geography of Difference* (1996) corrects this oversight, relying heavily on Raymond Williams' materialist cultural theory. Harvey's starting concept is the thorough imbrication of humans and nature: the "insistence that social beings can never escape their embeddedness in the world of nature and that no conception of political action could, in the final analysis, afford abstractions that did not encompass the fact" (1996:26). Harvey then quotes Williams: "Once we begin to speak of men mixing their labour with the earth, we are in a whole world of new relations between man and nature and to separate natural history from social history becomes extremely problematic." It is labor, then, that is the crucial point of entry into the interplay of understanding nature, history, geography, and society. This interrelationship is a dialectical one (1996:174–5):

> If . . . all socio-political projects are ecological projects, and vice versa, then some conception of 'nature' and of 'environment' is omnipresent in everything we say and do. If, furthermore, concepts, discourses, and

> theories can operate, when internalized in socio-ecological practices and actions, as 'material forces' that shape history, then the present battles being waged over the concepts of 'nature' and of 'environment' are of immense importance . . . What this in effect means, is that dominant systems of power can advance and protect a hegemonic discourse of efficient and rational environmental management and resource allocation for capital accumulation.

Discourses regarding nature and environment, then, have a role in the historical process—ideas operate in material ways when attached to practices. Such discourses on nature have helped to construct the environment as something that could be dominated, ultimately for control of the population and the territory, but also for the accumulation of capital—an important goal if the state is to maintain legitimacy (O'Connor 1973). Furthermore, these discourses, if hegemonic, are based in dominant systems of power and thus supercede and/or absorb opposition. The dams investigated in the second half of this work offer a set of case studies where one can see the state, through the building of an infrastructure, enabling continued capital accumulation. This accumulation is, then, based on the domination of nature, the absorption of opposition, and the discourse of progress, efficiency, and individual frontier heroism.

A discursive analysis of this process (state building so as to control population, territory, and enable accumulation of capital) must focus on the practices of the various groups of actors. State officials, local and regional elites, the press, and activist groups all contributed to the discourse about the building of a water infrastructure by the state in the American West. These multi-vocal contributions show a strong civil society and give weight to the argument against high modernism as inevitably leading to the overextension of state action. Instead, in one case it was the civil society groups who pushed the state to act, while in another case, activist groups kept the state from building a set of dams.

The deep analysis of these historical discourses regarding the dams of the West in the next chapters will show the importance of cultural constructions of nature as they were connected to material practices and power—constructions of water as something to be put to use, and of a river as something to be dominated. As it is expressed through such discourses, the political and cultural practices of expanding an economy, building state institutions, and engineering actual structures, show a particular understanding of the relationship between nature and society. While this relationship was understood as simple and direct, it is, in reality, complex and dialectical.

One example of this complexly shifting relationship is that of class-based relations between nature and humans over the middle decades of the twentieth century. While 1920s ideas of nature tended towards an anthropocentric appropriation of natural resources for human use,[10] in the '30s and '40s there was a shift towards greater conservation practices, as evidenced by the rise of New Deal land programs. As will be discussed in detail in chapters four and five, views of nature were not greatly differentiated by class; farmers, urban working classes, indigenous groups, and elites all supported the building of dams in the 1930s. There was no vocal opposition to Boulder and Grand Coulee Dams.

One of the main shifts in the relationship between nature and society that this study highlights is the rise of an effective oppositional group. As will be discussed in detail in chapter six, in the 1950s the Sierra Club and other environmental groups forged a coalition to oppose the incursion by the Bureau of Reclamation into Dinosaur National Monument. Opposition was spearheaded largely by several individuals (David Brower, Wallace Stegner, Bernard DeVoto), but their ability to vocalize the complaints of a large portion of the population disproves the "great man of history" approach. The Sierra Club and other organizations had the support of a burgeoning middle class, many of whose members valued nature as a place to escape from the rigors of the workplace, as a realm of recreation and relaxation, and as a place that had inherent value (Cronon 1996: 78; Gottlieb 1993: 41, 311).

The rise of the American middle class occurred in tandem with several changes in the structure of industry and society. Shifts in occupational structure (primarily that which included women in previously excluded jobs), shifts in industry and labor processes (the expanded importance of defense, technology, and the destruction of skilled labor), and alliances between previously isolated elite groups (Eisenhower's famous military industrial complex) provided the opportunity for the formation of an American middle class with its own social identity (Baxandall and Gordon 1995; Braverman 1974; Burawoy 1982; Lotchin 1992; Mills 1951, 1956).

This middle class identity was partially predicated on the increased emphasis on the nuclear family as the core unit of society. This tendency was reinforced by the white middle class relocation to the suburbs from the urban center, with a mounting reliance on the automobile for commuting, and the construction of multiple centers of commerce to supply the suburbs (Jackson 1985). One of the unintended, and perhaps ironic, results of this process was a greater attention to nature as a place of escape.[11]

Changes in infrastructure such as the proliferation of cars and the construction of the Interstate Highway System increased the rates of suburbanization, propelled the decentralization of factories and businesses (Jackson 1985: 265–66), and allowed for greater access to nature for the middle classes. Though roads connecting the National Parks had been mapped out in the 1920s, many roads to the Parks and within them remained in poor shape until the 1940s (Flink 1988: 173). A boom in car sales, however, matched the boom in baby making after World War II as civilian automobile production resumed after the wartime hiatus (Flink 1988: 190). Travel magazines, tourist organizations, and voluntary associations like the American Automobile Association promoted National Parks and Monuments as vacation destinations, and Americans visited wilderness in record numbers. Some publicized tourist destinations even included the subjects of later chapters: Boulder and Grand Coulee Dams. The 1947 Norman Rockwell painting "Going and Coming" shows a family traveling, one imagines, to a scenic location for a weekend getaway in their station wagon. The painting demonstrates the extent to which, after World War II, middle classes began to incorporate nature into their lives; one result of this was an increased value placed on nature and wilderness in the political sphere. The class basis of the politics of wilderness preservation is thus part of the complex evolution of discourses about the role of humans and their environment.[12]

This dialectical relationship between nature and society is also apparent in the process of state building in the American West. In order to take advantage of fertile soils and a long growing season, a plentiful supply of water had to be secured. This supply had to eventually come from the state, for individual capitalists could not make a profit from such a large and long-term investment. By combining the perspectives of environmental and political sociology with those of geography and history, we can make sense of the complex ways in which the state imposed itself on the landscape in the American West.

CONCLUSION: THE STATE OF NATURE AND THE NATURE OF THE STATE, A DIALECTICAL THEORY OF NATURE AND THE STATE

Rather than reducing the processes of state building to one of a number of factors—American exceptionalism, global competition, rapid industrialization—the dialectical view of nature and society promulgated here offers an understanding of state building in the American West as complex and contingent. Rapid industrialization and population growth in the eastern U.S. and global competition certainly matters to the history of the West (imperialistic chaffing with Britain, Russia, France and Mexico all contributed to the development of the frontier and the specific parameters of the U.S. boundary). But the history of securing a firm boundary through territorialization proceeds further than the Treaty of Guadalupe-Hidalgo in 1848 and the relinquishing of British claim to Oregon territory in 1846. State building processes—the widening of state capacity for economic and population control—continued well into the twentieth century as the state constructed an infrastructure on the landscape of the West. This achieved several goals of the state that ensure legitimacy: social control (Pisani 1983), territorial bounding (Scott 1998), and support for the continued accumulation of capital (O'Connor 1973) through a spatial fix (Harvey 2000).

State theory has historically not been "green" enough, while environmental theory has not paid enough attention to the interactions of the state and civil society. State-building mechanisms had to confront nature in the American West. Such mechanisms included the building of an infrastructure to overcome the vastness of the West and the environmental barriers to the continued accumulation of capital. The American expansion of administrative capacity that Skowronek (1982) described for the east were extended to include the administrative ordering of nature and territorial bounding (Scott 1998) that also offered a spatial fix for capital (Harvey 2000). This is to say, both the nature of the state and the state of nature are important to understanding the political economy of the American West.

The only analytic sufficient for encompassing spatial, temporal, cultural, political-economic, and environmental factors is a dialectical conception of society and nature. As Harvey (1996: 175) has argued, all socio-political endeavors are ecological on some level. We should add that they are all cultural on some level as well, and these endeavors also exist within both space and time. A dialectic of society and nature offers an analytical frame for understanding actions as disparate as dam building or the enclosure of lands in the name of the public domain. The dialectic of nature and society will be exemplified throughout the empirical chapters, but it can be theoretically characterized using the example of the epigram to this chapter, Woody Guthrie's lyrics to "This Land Is Your Land."

The lyrics to the song show a set of assumptions about citizenship, property, and nature. The land, according to Guthrie, belongs to "you and me" through our imagined community of citizenship in a bounded nation-state, but this is also in partial contradiction to normal American views regarding private property. Furthermore, the land was "made" by nature expressly for us, the citizens. While Guthrie both exemplifies and reproduces the cultural constructions of the landscape, the constructions also presuppose a set of state institutions and actions. As a dominating social and political power, the state has shaped the American West ideologically by imposing its way of "seeing" on a frontier territory, and materially by constructing space through an abstract mapping of the landscape and by building an infrastructure that enabled capital accumulation and population expansion.

The dialectic of society and nature also shows how the state has been shaped by history and the environment. Large new bureaucracies were created or reshaped to remove indigenous peoples and to enforce the administrative ordering of nature, while a growing technological elite runs the infrastructure and the economy, and a stolid middle class consumes an imagined wilderness. This wilderness is enjoyed by a privileged few while Americans in general continue unsustainable urban and rural resource-use patterns that endanger the longevity of their cherished wild places. A dialectical analysis of nature and society can take into account the disparate actions of the state, civil society, cultural practices, and the discourses on nature and conceptions of the use of nature by society.

Such a dialectical analysis must, however, be deeper than it has often been (Harvey 1996: 183) and connect with actual material practices rather than remain on a discursive level. Ideas and discourse matter only as they become connected to powerful material practices and processes that can have real effects. Thus, the actions of the state, based on certain cultural conceptions of nature, have the power to shape nature; and yet the ways in which nature exists (geography, climate, etc) and how nature reacts to societal actions can similarly shape the role of the state, civil society, and economy. A dialectical analysis of these processes is the task of the empirical work of this book, and the three empirical chapters will thread through the historical discourses about water systems in the American West. The chapter that follows shows the overarching historical frame.

Chapter Three

Water in the American West

> You just watch this river and pretty soon
> everybody's gonna be changing their tune.
> The big Grand Coulee and the Bonneville dam
> are running a thousand factories for Uncle Sam
> and everybody else in the world.
> Making everything from sewing machines
> to fertilizer . . . atomic bedrooms . . .
> Plastic! . . . Everything's gonna be made out of plastic.
>
> Yeah, Uncle Sam needs wool, Uncle Sam needs wheat
> Uncle Sam needs house and stuff to eat.
> Uncle Sam needs water and power dams,
> Uncle Sam needs people . . . and people need land.
> Don't like dictators none-much myself,
> but I think the whole country should be run by . . .
> Electricity!
>
> "Talking Columbia," Woody Guthrie

The history of the relationship of the American West to water scarcity is expansive. The goal of this chapter is to parse this history by focusing on the changing ways that humans have dealt with resource scarcity in the West, and how these actions conditioned the nature-society dialectic over time. I will pay special attention to the changing role of the state in this complex and dynamic process. Much of the bureaucratic infrastructure and physical building was done during the Progressive Era and the New Deal, so this chapter focuses on these eras in particular detail.

The issue of water scarcity in the American West is an important topic for both present policy and the history of the region in several respects. First, the construction of massive dams and irrigation projects is the most material outcome of an era of progressive government. The progressive manifestation of the utilitarian ideology of "the greatest good for the greatest number," as it was institutionalized in the Bureau of Reclamation and other Progressive-Era bureaus and offices, represents an acceleration of bureaucratic rationalization and human control of nature. Second, the establishment of large-scale water systems had profound effects upon the environment and society of the region. The large-scale water systems, in conjunction with other social and historical factors, have helped bring the region to national prominence, especially the state of California—now the most populated and productive state in the U.S. Third, this government-induced growth model has been consciously mimicked internationally, and so the American West has been a model for global development.

Human transformations of the land also have historical significance as indicators of how humans view nature; the large-scale manipulations of the West, such as dams and canals, expose an ideology of domination (Worster 1990). Although humans have often imagined themselves as separate from the ecological systems in which they live, and from which they draw their sustenance, humans have always managed to turn nature to economic ends, and in so doing have transformed their immediate surroundings, often with unintended consequences (Cronon 1983). A historical analysis of these manipulations of nature can be a window into understanding the motives, methods, and consequences of how humans view nature.

In the American West, water sources were a prerequisite for frontier expansion by Anglo settlers, and much of this process culminated in the Progressive Era. With the rise of Progressivism came the zenith of the ideology of domination, in which humans believed that they could scientifically manipulate nature in predictable and understandable ways. For the Progressives, these alterations were intended to produce a society where large amounts of people gained from the benevolence of the government and nature. The Progressives were not entirely wrong in their predictions; the West has indeed benefited economically from the actions of the government in the Progressive Era. But unintended consequences have also ensued from these transformations of the environment.

The regional economy would not be what it is today without the large-scale water systems built in the Progressive Era and the New Deal, when the government greatly expanded its conservation capacity. The powerhouse of the West's economy is clearly California, which if viewed on its own is ranked as the sixth largest economy in the world (Ca.L.A.O. 2000: 1). A variety of industries contribute to California's economic dominance: agribusiness, high technology, the defense industry, and culture production in Hollywood. Dependable water sources are crucial for all of these sectors of the economy;

they sustain workers and local consumers and they are required if California is to compete in national and global markets. California fruits and vegetables are eaten internationally, its movies are watched around the world, and the technology produced in Silicon Valley has enabled the newest form of the global economy. The defense industry, in particular, is directly dependent upon federal water projects. Cheap hydroelectric energy helped spur the growth of what Lotchin calls California's metropolitan-military complex (1992). But such paths to prosperity have been uneven over the West—some less powerful states have competed with California over resources and political capital, not always winning, but also not always losing.

HISTORIOGRAPHY OF WATER AND THE AMERICAN WEST

Water is crucial to California's dominance, but much of the early reclamation historiography tended to romanticize western development (Lee 1980) and not recognize the complexity of nature-society relationships. On the contrary, works since the late 1970's have attempted to establish both the continuities and disjunctures between past and present human transformations of the land. Using classic works such as Walter Prescott Webb (1931/1959) as a guide, the latest wave of reclamation scholarship provides a rich, nuanced, and synthetic view of the history of the American West. This historiography focuses on three main issues: the history of California, California's uneasy relationship to the other states of the West, and the influence of agriculture and mining in the development of water rights.

Norris Hundley (1972; 1973; 1975; 1992) has shown the ways in which water, land policy, and law have shaped the development of California and the West. His 1992 work, *The Great Thirst* shows the ways that indigenous people in the southern California region shaped their environment in order to overcome the problems of aridity, and narrates the coming of the Europeans and the development of the American state.

Hundley notes that the Spanish wrote the first conscious water policy in the region. The Spaniards built their first permanent settlements in southern California in 1769, located near water sources and indigenous populations. The "law of the pueblo" commanded that upstream users could not restrict the water access of a settlement downstream. This policy was based on the needs of the community instead of those of the individual and was tailored to the rigors of life on the frontier. The Mexican interlude, starting in 1821, provided little change for the water policy of this area (Hundley 1992: 42).

Hundley also shows, however, that the American victory in the Mexican-American war in 1846 did force some major changes in water law. One important distinction between the Americans who overran the California region and the former Spanish governors was the American's individualist and market-oriented world-view. Nature was viewed as a source of wealth rather than something to be respected or used wisely for the community (Hundley 1992: 44).

The Gold Rush brought several major changes to the region just after the defeat of Mexico. With only a local governmental presence, what rules there were received lax enforcement. And when combined with the individualist orientation of miners and other profiteers, it produced a reorganization of the Spanish community water policy to one of "first in time, first in right." This evolved into what is commonly known as the prior appropriation law (Hundley 1992: 41, 45, 62). This outgrowth of the mining ethic posits possession as the greater part of the law. That is, the first person to use the water had the rights to it.

The evolution of water rights in the West was a complex process (Pisani 1992; Robbins 1976). From the origins of prior appropriation, Hundley (1992: 67–101) traces the development of the "California doctrine," a set of laws and precedents that established a balance between prior appropriation claim holders and the traditional eastern water policy of riparian rights. Descended from English common law, riparian rights make little separation between rights of a landowner to use *land* and the rights of a landowner to use *water* adjacent to that land. A river that flows through a property could be used by the landowner without limit. In an arid land, neither the riparian policy nor the prior appropriation policy makes much sense, and the California doctrine blended the two traditions into an ungainly synthesis. The California doctrine arose from the *Lux v. Haggin* (1886) decision, in which both prior rights and riparian rights were recognized, depending upon when the riparian rights were purchased. Hundley argues that this judicial compromise has led to a controlled legal chaos in many states of the West. A water user could sue if s/he found someone infringing upon his or her rights, but every such downstream claim must be filed separately, making water rights a perpetual problem in the courts (Hundley 1992: 97).

Donald Pisani (1992) identifies an alternative path in Wyoming water policy. Wyoming created an administrative—rather than judicial—solution to problems of water rights. The state of Wyoming declared, in its 1889 constitution, that all water within the state was state property, and was to be used for the public interest. Wyoming defined water rights in such a way as to give state technicians complete control over distribution—including control over appropriations made prior to the passing of the state constitution. This administrative solution was used by a few other western states and represented a move towards collective control by experts rather than laissez-faire policies and judicial precedent (Pisani 1992). Control by elites in Wyoming in some ways anticipated what was to come with federal technocratic control by the Bureau of Reclamation.

Differences in water policy between the various states of the West have had varying consequences for development, and reflect the power of different interests in each state. In California, Hundley argues, the power of cities, landowners, and miners all influenced the development of a water policy that had an *ad hoc* character. The California doctrine has led to speculation and fraud (Hundley 1992; McWilliams 1949). The existence of the Wyoming doctrine, Pisani argues, is an indication of the fragmentation of the West. States had radically different policies, each in its own way attempting to avoid federal administration of public lands and water in the West (Pisani 1992: 68).

This fragmentation between states of the American West was represented by differences in wealth, population, and industrial and agricultural productivity. Geographical differences also increased the power of upstream communities over those downstream. In the fights over rights to interstate rivers, the federal government was expected to be a key player. Its specific role, however, was contested until the development of the Colorado River Compact Commission in 1922, in which backers of the Boulder Dam Project had to navigate through the morass of interstate politics and long-time western rivalries, eventually producing the Colorado River Compact (Hundley1975).

The compact represented several significant changes in the federal presence in the West (Abbott 1994). First, the compact showed the potential for federal intervention, for it provided an unprecedented compromise between states on the complicated issue of water law. Second, the compact initiated a national debate over hydroelectric power, regarding whether it was more appropriately administered by private or public agencies; and if public agencies, then what level of government would regulate (federal, state, or local). Third, the compact made possible a series of large dams, starting with the Boulder Canyon Project. The landscape of the West would never be the same.

PROGRESSIVISM, CONSERVATION, AND WESTERN HISTORY

Similar to the fragmentation in the historiography of water and the west, despite the best analytical efforts of historians, the definition of Progressivism remains contested. Daniel Rodgers argues that Progressivism was a large collection of diverse social problems and political solutions that are best understood in light of the "movements of politics and ideas throughout the North Atlantic world that trade and capitalism had tied together" (Rodgers

1998: 3). While this expansive understanding is certainly true, such an analysis is too broad to be useful for local studies. Since Progressivism has complex and contradictory aspects to it, perhaps it needs to be broken down into parts to be understood. While it was not a unitary phenomenon, many historians find dichotomous factions in the Progressive movement (Buenker 1973; Hackney 1969; Holli 1969; Thelen 1976).

More persuasively, according to Arthur Link and Richard McCormick (1983), we can explicitly identify three strands of Progressive thought: 1) social progressives, 2) reforming professionals, and 3) coercive progressives. These three categories loosely map onto Daniel Rodgers's identification of the central Progressive concerns: 1) social bonds, 2) social efficiency, and 3) anti-monopoly (Rodgers 1982). Though these three strands sometimes overlapped, they often represented quite distinct immediate goals, diverse methods, and multiple motivations. At times these three social groups contradicted one another and confused the term Progressivism almost to the point of incoherence. As McCormick has mentioned, the early twentieth-century Progressives manifested a "baffling diversity" (McCormick 1986). Yet Progressivism as a term continues to be used both in popular discourse and scholarly works. It is a worthwhile endeavor to clarify it. McCormick shows that these three loose-knit groups shared a core set of abstract ideas related to purpose, rationale, and results.

In the last two decades of the nineteenth century, America experienced a profound increase in industrial growth. Fueled by capital investment from European banks and presided over by financial innovators such as J. P. Morgan, American industry expanded and concentrated. Famous entrepreneurs like Andrew Carnegie and John D. Rockefeller made fortunes in sectors such as manufacturing and oil. Progressives shared a common understanding (and embrace) of this process of industrialization. They did not, however, accept the corruption of business and politics that had come with it (Robbins 1976: 302). This suspicion of corruption was addressed by positive action that was aimed at improving human communities through direct intervention in social activity. This action was, in part, based on the research undertaken by reformers. Using the methods of the new social sciences as their model, Progressives attempted to understand society, and thereby change it.

Progressivism as a self-descriptive term was not used until about 1910. By this time, however, many of the social movements that labeled themselves as Progressive had been operating for twenty years or more. They had acted in a wide range of locales and with many goals. They established settlement houses in run-down neighborhoods, fought sin and vice, educated immigrants, and lobbied for anti-trust legislation. They reformed government to rid it of corruption, and held up the model of the professional, objective bureaucrat. As discussed in chapter two, many of these elements—such as the emphasis on social power belonging to the office rather than the individual—brought the U.S. into modernity. Still other elements, such as a strong religious morality, cannot be easily characterized as modernist. Though these elements were at times contradictory, it was through them that the push for progressive reform was felt both on Capitol Hill and in local halls of governance. New politicians and administrators from diverse states went to Washington, D.C. and helped to bring about what Robert Wiebe has called the bureaucratic revolution (Wiebe 1967).

The new social sciences contributed to the allure of rationalization. In a series of essays from 1904–1919, Max Weber argued that science could help to increase knowledge about society, especially in regards to the most efficient and practical ways to achieve certain ends. Weber, however, also asserted that science could not help society adduce what such ends should be (Weber 1922/1946: 143). Sociologists such as Auguste Comte and Emile Durkheim similarly claimed that society could not only be understood, but also could be altered, much like nature or industry (Mandelbaum 1971: 237; Manuel 1962: 278–9). This crucial aspect of the ideology of modernism—the human mastery over nature—was extended to human mastery over society as well. No longer would humans have to suffer from the vagaries of filth, corruption, and inefficiency. Nature, both human and nonhuman, could be controlled by society through the proper application of rationality. As Scott notes in his definition of high modernism (1998: 89), the Progressive Era exhibited "a supreme self-confidence about continued linear progress, the development of scientific and technical knowledge, the expansion of production, the rational design of social order, the growing satisfaction of human needs, and, not least, an increasing control over nature (including human nature) commensurate with scientific understanding of natural laws."

In the American West, this set of Progressive ideologies combined with new ideas of conservation to produce Theodore Roosevelt's popular restatement of the utilitarian dictum "the greatest good for the greatest number for the longest time." Roosevelt's twin conservation policies of reclamation and forest conservation worked from this utilitarian idea and brought together the variegated ideologies of Progressivism. The 1902 Reclamation Act financed irrigation projects through the sale of public lands in the West. The Act gave the Secretary of Interior and selected experts complete authority over projects, whose goal was development of the West for the social good

(Pisani 1983). Forest conservationists acted similarly, and the two conservation constituencies worked together under the motto "Save the Forests, Store the Floods, Make Homes on the Land" (quoted in Hays 1969: 26). Together, they helped solve a persistent problem of the frontier: how to secure the apparent natural abundance of the West.

Writing in 1964, Samuel Haber suggested that the Progressive Era was the time when there was the closest linguistic match between the concepts of "efficiency" and "good." Haber's study focuses closely on Frederick W. Taylor, who, in 1895, invented the concept of scientific management. Taylorism was essentially a process of labor control in the factory. A piece-rate system was instituted to give an incentive to workers, thereby making them more efficient. Taylorism was supposed to bring worker's interests into line with management interests; with the piece-rate, both groups had a financial stake in speed, perfection, and output.

Taylorism became connected to Progressivism through an odd affinity. Though Taylor was focused primarily on labor control, the Progressives were interested in the abstract concepts of increasing human potential. This tension, like many of the contradictions within the progressive movement, was not troubling to proponents of Taylorism. As Haber describes it, "In most cases, however, [scientific management] was spread not through conversion but through something more akin to syncretism. Scientific management was in the public eye, and leaders of other causes dependent upon popular attention began to translate their appeal in its terms. This was most apparent in the conservation campaign" (Haber 1964: 61).

Conservation was originally a set of laws and policies that arose out of the Progressive impulse to curb the power and destruction of the corporations. "Not until the railroad magnate, the cattle king, the mining baron, and the lumber monarch had established a prestige as great as that enjoyed by the capitalist of the eastern industrial order, did the federal government finally pass the first of a series of laws which was ultimately to be distinguished as the conservation movement" (Robbins 1976: 301). The Progressive concerns with corruption and efficiency articulated with the need for control over the territory on the level of mapping and classification in the report of the Public Land Commission of 1880. Robbins suggests that this report had three major results: "first a realization that fraud and corruption existed under the prevailing system; second, a recognition of the urgent need for reform and codification of the land laws; third, the conviction that thorough survey and classification of the remaining lands, and new legislation to govern their disposition, were highly imperative" (1976: 302).

The conservation movement was also a prime example of the governmental changes wrought by Progressivism (Hays 1969). The new forms of political organization instituted by the reformers had far-reaching implications for the way that decisions were made and the type of democracy that was practiced in America. The new forms of organization had the effect of moving decision-making processes away from the grass roots to the realm of the elite technician. Hays called this the "upward shift." This shift was a new form of social control, one that was a "more elaborate process, involving measurement and prediction, reliance upon the experts who could develop and manipulate information, and techniques for shaping the course of events to reach predictable outcomes" (Hays 1969 unnumbered preface page).

Through the syncretism of efficiency and conservation, the logic of the state in the Progressive Era came to reflect ideals of rationalization. For the Progressives, the proper role of the state was to build the necessary infrastructure for the development of the economy. This period of federal involvement—the conservation movement—stems from the Forest Reserves Act of 1891, and the creation of the Reclamation Service by the Congress in 1902.

Preservation or conservation?

The Progressive Era was, thus, rife with contradictions and vagaries, and Progressive-Era conservation thought was no different. In fact, during this period there developed a split among those who believed in a need for the conservation of resources. On one side were Gifford Pinchot and the majority of government administrators, who held that resources must be used wisely in order to conserve them for the future (Pinchot 1968). On the other side were those who believed in the preservation of wilderness based on aesthetic and spiritual values that, they argued, superceded all other human uses. This viewpoint is most closely associated with John Muir and his associates in the Sierra Club (Teale 1954).

Progressive Era conservation thought was heavily influenced by several key works of the late nineteenth century. In 1864 George Perkins Marsh published *Man and Nature,* arguing that "Man was given nature for usufruct alone, not for consumption, still less for profligate waste" (quoted in Nash 1967/1982: 105). Furthermore, Marsh saw that conserving nature was not just compatible with, but necessary for economic progress. Conservation thus had, as Nash indicated "'economical' as well as 'poetical' justifications" (Nash 1967/1982: 105). This dual construction—economic and aesthetic—prefigured the conservation/preservation conflict in the Progressive Era.

Frederick Jackson Turner, in his famous frontier thesis of 1893, argued that in addition to Marsh's two principles, a third was needed to explain the value of the American frontier (1920/1996). In its harsh demands on pioneers, the frontier, Turner argued, was the source of the peculiar style of American democracy. With the closing of the frontier, went Turner's story, democracy was threatened. Turner's and Marsh's works gave expression to a new anxiety for Americans; as Cronon has argued, Americans began "fearing the end of abundance" (Cronon 1994: 606).

The federal government responded to such fears by mapping and protecting the landscape. The 1891 Forest Reserves Act was the first of the Progressive-Era conservation laws. Many other laws, executive orders, and administration policies were written and enforced, including the 1902 Reclamation Act and the creation of national parks and monuments. These laws and policies were intended to protect resources from overuse through rational management techniques, and also served to further the control of the state over the territory of the West.

Such laws and policies, however, ran aground on the divergent values that contributed to the fear of scarcity. Were the forests put under reserve for "wise use" as Pinchot argued? Or were they protected for the "poetical" reasons of Marsh and Muir? Should the new lands opened by the Reclamation Act be reserved for the small family farmer to protect the democratic spirit or should they be distributed based on the "greatest good for the greatest number," favoring the (presumed) efficiency of the larger corporate farms? These questions came to pointed dispute in the Hetch Hetchy controversy of 1905–1913, in which John Muir battled Pinchot and Roosevelt for what Muir perceived to be the soul of the West: would a dam be allowed in Yosemite National Park to slake the thirst of San Francisco?

San Francisco proclaimed the need for a dam at Hetch Hetchy Valley as early as 1905, and redoubled its efforts after the earthquake and fire of 1906 (Clements 1979). Hetch Hetchy Valley is part of Yosemite National Park, and thus the city needed approval from the White House to build the dam. Muir and fellow Sierra Club member Robert Underwood Johnson argued that despoiling the natural beauty of Hetch Hetchy Valley was akin to desecrating a cathedral, and the dam should be built elsewhere (Johnson 1968; Teale 1954). Gifford Pinchot, Interior Secretary James R. Garfield, and others in the Roosevelt administration were for the dam. President Roosevelt himself, however, was ambivalent. Having camped in the region with John Muir a few years earlier, Roosevelt recognized the splendor that would be drowned. But the utilitarian in him saw the needs for water, for population growth, and for the wise use of resources (Richardson 1959).

Roosevelt eventually decided to support the dam, but insisted that Congress had the final authority.

The battle thus moved to Congress, where Muir was sure that the people of America could have their voices heard. The Sierra Club and several periodicals started a letter-writing campaign that was unprecedented for the time—one senator received more than 5,000 letters (Nash 1967/1982). The bill in the House, however, turned on the support of the Public Lands Committee.

William Kent was a key member of this committee, and an ardent reformer in the Progressive spirit. Before the Hetch Hetchy struggle, Kent had purchased a tract of land in his home-district of Marin County (just north of San Francisco). This land was virgin old-growth redwood forest, and he gave it to the United States with explicit instructions to preserve it as a National Monument under the name of one of his idols, John Muir; thus Muir Woods was created. With such an ally in the House, Muir was sure that the bill could be killed (Nash 1967/1982).

Kent, however, was not only an admirer of Muir, he was a detractor of private electric power. A dam at Hetch Hetchy would produce publicly-owned hydropower, which would help to defeat the growing monopoly of Pacific Gas and Electric. Kent used his reputation as a preservationist to support the Hetch Hetchy dam, and it eventually passed in the House and Senate in 1913. With few more perturbations, the dam was built, and Hetch Hetchy Valley today is a slack-water reservoir (Clements 1979; Nash 1967/1982).

In a sense, then, conservation rather than preservation principles won out in the late Progressive Era. The poetical, spiritual, and aesthetic values inscribed onto nature by the preservationists were out-argued by the utilitarian logic of the conservationists: Hetch Hetchy was built in part to satisfy the greatest good for the greatest number, and in part to fight the corrosive power of the corporations.

Dating the end of the Progressive Era

Most historians date the end of the Progressive Era sometime between the end of the popular reform movements in 1917 and the beginning of the New Deal (Link and McCormick 1983). The 1920s certainly did not simply continue the policies and practices of the Progressive Era, nor could they be said to prefigure the New Deal. The presidential administrations of Warren Harding, Calvin Coolidge, and Herbert Hoover saw very little continuation of Progressive Era conservation policies; this would have to wait until Roosevelt's New Deal, when the President put in place a new set of programs that built from many of the ideas of the Progressive Era. Hoover is perhaps the only exception to this trend, for he made oil conservation, navigation, and flood control part of his administration's goals. Hoover also was important—both before and during his presidency—in enabling the Boulder Canyon Project, among other conservation-minded programs (Swain 1963).

Though the decade of the 1920s may have been more subdued in conservation thought and practice than the years immediately preceding and following it, there are, in fact, only slight differences between the Progressive Era and the New Deal in terms of the projects proposed for the West. Many of the same actors continued their work seamlessly through changes in administration, and as Worster (1994: 271) points out, published "culminating works" during the New Deal. Many of the governmental programs established in the Progressive Era were not completed until many years later. For example, the precursor to the Bureau of Reclamation was created in 1902 but it did not reach the height of its power until the completion of the Boulder Canyon Project in 1935. Grand Coulee dam was first proposed in 1918, but was not completed until 1941. Though many Bureau projects that were initiated during the Progressive Era were not completed until the middle of the Great Depression, these projects still reflect the spirit of Progressivism due to the time of their conception and their ideology of efficiency, progress, and faith in technology. As Kathy Morse has noted, New Deal conservation is in some ways Progressive conservation with a whole lot more money (Morse 1999).

THE NEW DEAL AND CONSERVATION

Conservation in the New Deal was, however, a bit more than just Progressivism with a fat wallet. The ways in which the government spent money on conservation was both more creative and more effective. Experimenting with using government spending to boost the economy, the Roosevelt administration often chose conservation efforts as the recipient: soil conservation, tree-planting, dam-building, and more. By spending on conservation practices throughout the nation, just as he had earlier as Governor of New York, Roosevelt could help revive the economy while avoiding competition with private capital (Owen 1983).

Perhaps the most creative aspect of Roosevelt's conservation program was its coordination. Roosevelt's plans for conservation of forests, soil, water, and energy were coordinated with private capital, between government agencies, and across different resources. "We seek to use our natural resources not as a thing apart," Roosevelt said, "but as something that is interwoven with industry, labor, finance, taxation, agriculture, homes, recreation, good citizenship. The results of this interweaving will have a greater influence on the future American standard of living than the rest of our economics put together" (quoted in Owen 1983: 84).

Conservation during the New Deal was thus a complex coordination of programs, policies, and appropriations. As one of his first projects, Roosevelt created the Civilian Conservation Corps (CCC), which was in many ways the prototypical New Deal conservation agency. The CCC, along with the Tennessee Valley Authority (TVA) and the Public Works Administration (PWA) were key parts of Roosevelt's relief and recovery programs (Owen 1983). Each of these agencies used conservation as a tool in the effort to revive the economy, re-employ people, and improve social life.

The relief/recovery/conservation programs put into place as the Roosevelt administration attempted to revive the economy by stimulating industry. The PWA was organized in June of 1933, with its clearest job being to "stimulate heavy industry by fostering public works that required huge quantities of material." This was a "pump-priming" attempt to stimulate private industry through compensatory spending. Much of the PWA efforts were aimed at water projects in the West, including spending on Grand Coulee and Boulder Dam projects (Owen 1983: 84).

Similarly, the CCC was meant to put unemployed Americans, especially youth, to work in "healthful surroundings" in order to "eliminate, to some extent at least, the threat that enforced idleness brings to spiritual and moral stability" (F. D. Roosevelt quoted in Nixon 1972: 144). The CCC, according to Roosevelt's speech to Congress, was to be used "in simple work, not interfering with normal employment, and confining itself to forestry, the prevention of soil erosion, flood control and similar projects. . . . This enterprise . . . will conserve our precious natural resources. It will pay dividends to the present and future generations" (quoted in Nixon 1972: 143). Like the PWA, the CCC stimulated the economy by employing workers on projects that were not competitive with private capital, but instead could boost the morale of workers while fending off the evils that (they thought) were brought on by unemployment. In this way the CCC acted as a social and moral force as well as an economic one, an aim of New Deal programs that was to find its full expression in the TVA.

The TVA, authorized in May of 1933, was given broad authority to improve the social, economic, water, and mineral resources of the Tennessee Valley. Roosevelt's project, initiated by Senator George Norris of Nebraska, was nothing less than an attempt at full social engineering, whereby the economy and society of the Tennessee Valley would be modernized (Roosevelt 1938: 466).

In many of these New Deal programs, Roosevelt's main advisor and administrator was Harold Ickes (Lowitt 1993). Continuities between Progressivism and the New Deal can be seen in Ickes' self-presentation in Progressive and utilitarian terms: "The principle underlying true conservation is the protection, upbuilding and prudent use of our national resources for the greatest number of our people" (Owen 1983). Like the Progressives before him, Ickes was devoted to fighting political graft and to efficiency: he "had an aversion to using federal funds 'to hire grown men to chase tumbleweeds on windy days'" (Ickes as quoted in Owen 1983: 84).

The Progressive Era and the New Deal, then, have a great similarity and continuity on the level of the human relationship to nature. While there are important differences, the New Deal is in many ways Progressivism with a larger state budget. Roosevelt implemented ideas and practices that he conceived of as Governor of New York during the late 1920s, and many of the dams proposed for the West during the 1920s were finally built with financing from the New Deal.

PERIODIZING THE RECLAMATION OF THE WEST

While paying attention to both the continuities and conflicts between Progressive-Era and New Deal conservation thought, we can divide reclamation history into four periods. In his historiographic essay, Lawrence Lee identifies three phases of the conservation movement (Lee 1980). I argue that a fourth period is required to make sense of the recent changes in reclamation policy.

The first phase of reclamation began with the creation of the Reclamation Service (later renamed the Bureau of Reclamation) in 1902, and the era continued until Boulder Dam was built. Over a period of 30 years, the Bureau proposed and built a series of dams and irrigation projects. Commissioners Frederick H. Newell, Arthur Powell Davis and Elwood Mead saw the Bureau through difficult times as the fledgling bureaucracy battled with farmers on the one side and Congress on the other. This nascent period of dam building is best characterized as 'reclamation under Progressivism.' The dams tended to be small in scale as compared with later projects, and they were generally single-purpose.

The Laguna Dam on the lower Colorado River characterizes this period well. Proposed in 1904 and completed in 1909, Laguna Dam is part of the Yuma Project, which delivers water to the Imperial Valley of California and Gila River Valley of Arizona. Relatively modest in the number of acres that the dam "reclaimed," the Yuma Project was expanded after 20 years of political and technical change. After botching several attempts to appropriate Indian Reservation land for (Anglo) settler use, the Bureau of Reclamation turned most of the water west to California. As technical knowledge improved, the Yuma Project and Laguna Dam were incorporated into the large-scale vision of Arthur Powell Davis as part of the Boulder Canyon Project, which characterizes the second phase of reclamation.

The second phase of reclamation begins with the completion of Boulder Dam in 1934 and runs through the completion of Glen Canyon Dam in 1963. Lawrence Lee thus suggested that the second phase of the reclamation started with the Great Depression and the New Deal, Marc Reisner calls these the "go-go years" (Reisner 1986: 145). Bureau Commissioner Elwood Mead, following the early lead of Commissioner A. P. Davis, pushed the Bureau towards large-scale and multi-purpose projects, the prime example being Boulder Dam. These large projects required immense federal negotiation, seen in the Colorado River Compact.

The success of Boulder Canyon and a number of other projects led to the Bureau of Reclamation being viewed as "the establishment" in the 1960s (Lee 1980: 49). Robert Gottlieb summarized this establishment as the "Iron Triangle," made up of Congress, the water agencies, and local water industry groups (Gottlieb 1988). Environmentalists of the 1960s had found a perfect opponent who was despoiling natural beauty while supporting capitalist expansion and population growth, and so reclamation entered a third phase.

The protest movements of the 1960s attacked such establishment agencies with the rediscovered ideas of environmentalism and preservation. A series of congressional acts responded to these protests, and the Bureau of Reclamation was forced to conform to the mandates of the 1964 Wilderness Act and the Wild and Scenic Rivers Act of 1968 (Lee 1980: 52). This third phase could be aptly called 'reclamation under environmentalism.'

Glen Canyon Dam best characterizes the inception of this phase. As will be explored in detail in chapter six, in the mid-1950s, the Bureau of Reclamation wanted to use the entire flow of the Colorado River; Glen Canyon and several other dams were proposed in 1956 as part of the grand scheme. Through intense lobbying in Washington D.C. the Sierra Club won what it thought was a victory over the Bureau. David Brower and the Sierra Club managed to eliminate a dam at Dinosaur National Monument (Echo Park), but the deal allowed the dam at Glen Canyon to go forward, and Glen Canyon was completed in 1963. Environmentalists decried the deal as a sell out, and Brower acknowledged the mistake, vowing "to never sacrifice a scenic-resource again" (Gottlieb 1988: 54).

This third phase of the reclamation movement is now, I argue, giving way to a fourth: the privatization phase. The Bureau of Reclamation has officially recognized that it cannot build any new dams in the West. As one report states, "the demand for water continues to increase as the supply decreases" (United States Bureau of Reclamation 1996). As Elwood Mead (Commissioner of Reclamation during the New Deal) once quipped, if all of the rain that falls west of the 98th meridian could be captured and used, it still would not slake the thirst of the American West. Furthermore, "the justification for . . . building new facilities has, by and large, gone away" (Hess 1996). The sources of constraint upon dam building are many, not the least of which are the natural limits. In fact, Mead's fantasy is now approached in reality: all but a few major rivers in the American West are dammed (most of them multiple times); in California alone there are some 1200 major water projects.

Responding to the call for budget reduction (prompted by President Clinton's National Performance Review), the Bureau of Reclamation has instituted a new policy called "title transfers." A transfer of title entails the full transmission of responsibility, benefits, and liability from the federal government to individual persons, corporations, or associations. The new policy represents a shift in power from the technical experts of the government to a new set of local actors.

The transferred projects, as well as those that in the process of being transferred, have different purposes, recipients, and implications for the West. Some transfers will go to municipal water districts. Others will be transferred to local control for various purposes. For example, the city of Dickinson, North Dakota, has applied for a transfer of the Dickinson Dam due to its importance as a recreational area. In a majority of cases though, the beneficiary will be an irrigation district. Title transfers, then, are a decisive move towards privatization.

By privatizing large-scale water systems, the federal government is essentially undoing a century of state building in the West. Without control over water systems, the state will have a reduced ability to navigate future problems of water allocation. This is especially important in the context of the dire predictions involving increasing water salinity and the silting of dams. If the state does not have direct influence over water systems, it can only encourage changes in use patterns rather than change supply levels.

Another consequence will likely involve the maintenance costs as the dams age. As the dams become less and less useful, it is unlikely that private entities or small municipalities will be able to afford the massive price tag of de-silting a dam. As the dams fill with silt, the economic character of the dams will shift away from producing hydroelectric power and water storage. The only obvious use for large bodies of shallow water would seem to be recreation, and the destinations of most tourists are the majestic sites that the federal government is unlikely to privatize.

The history of water and the American West, vast as it is, shows the continuities and changes over time of the relationship of humans to their environment. Through the settlement of the western frontier, the American state addressed concerns about scarcity by building an infrastructure on the landscape. This infrastructure—the dams, canals, and power plants—has enabled the West to lead the U.S. in many respects. But this changing relationship between humans and nature is also evident in the discourse used during the debate over the dams. This is the subject of the next three empirical chapters: how was the nature-society relationship constructed, and how was this based in domination.

Chapter Four

In the Beginning There Was Boulder: A Natural Menace Becomes a Natural Resource

Abe Lincoln freed the Negroes
And old Nero he burned Rome,
But the Big Six helped depression
When they gave the stiff a home.
In a nice bunk house they're sleepin'
They're workin' every day,
The hungry look has vanished
For they got three squares a day.
You'll find tall Lou from Kalamazoo,
And Slim from Alabam,'
Mixed in with all the rest of us
Old boys on Boulder Dam.

And the fallin' rocks can't scare us
Nor the scorchin' rays of the sun,
We've rode the rods and brakebeams
Ragged and on the bum.
And they gave us jobs and fed us
When we needed it you bet,
And we all are truly thankful
With no feelin' of regret.
So we're stickin' till the finish
There's me and Ike and Sam,
And we're getting' fat and stakie
Us old boys on Boulder Dam.

"Old Boys on Boulder Dam" written by Claude Ramer, Boulder Dam worker, published in the *Las Vegas Evening Review-Journal,* 9 November 1931.

Abe Lincoln freed the Negroes,
Old Nero he burned Rome;
But the Big Six fills a graveyard
With the stiffs without a home.
In a bunkhouse barn we're sleeping
And toiling with might and main
But you'll never hear us crabbing
'Cause we hope to scab again—
So you'll find Shorty from Ossining
And Slim from Alabam'—
By golly, nearly all the crew are cussing
US OLD SCABS ON BOULDER DAM!

We don't mind the falling rocks
Nor the scorching rays of the sun
For our heads are solid ivory blocks
And we think its lots of fun
We were ragged, buzzin' in the jungles;
It's better than Sally soup, you bet
So we're all mighty humble
With not a feeling of regret
And we're sticking to the finish—
There's Woody, Mac and me! Damn
But we're getting fat and scaley
US OLD SCABS ON BOULDER DAM

"Old Scabs on Boulder Dam"
Published in the *Industrial Worker*
12 January 1932,
with "no apologies to Claude Ramer."

Intense battles marked the beginnings of the debates, the actual construction, and even the dedication of Boulder Dam. First proposed by Mark Rose and the Imperial Irrigation District in 1911, the dam was fought over by western states, debated by farmers, power companies, media moguls, Congress, and Bureau of Reclamation engineers. Finally approved in 1928, and constructed from 1931–36, Boulder Dam established the foundation for mid-century state-building discourses that were infused with the rhetoric of dominating nature and subordinating it to human ends. The Colorado River was variously described as a "tyrant," a "raging river," and a "natural menace." In order to overcome these natural barriers the Bureau of Reclamation, Congress, and several Presidents of the U.S. acted (sometimes in consort, sometimes at cross-purposes) to discursively construct the river as a "natural resource."

This conversion of natural menace to natural resource occurred through language as well as through the actual building of the dam in the river. The discursive construction of the Colorado River as a natural resource contained elements of appreciation for nature as a productive force as well as a deprecation of nature as "red in tooth and claw." There were also strong elements of state-building rhetoric that characterized the river as a potentially useful resource—a key element in the building of an empire in the American West. Using the words of the historical actors themselves, this discourse is best labeled *imperialistic.* Various social groups fought over how this empire was to be built—and who would benefit from the resource use. Private capital battled for control of the electricity, local farmers and their Congressional delegates fought for water rights, and the many Depression-struck jobless jockeyed for employment while union activists struggled to organize them. These social groups all engaged in discursive and sometimes physical battles over Boulder Dam.

Their discourses can be separated into two main categories; both of them are best labeled, using the terminology of the historical actors themselves, as imperialistic. The discourse of the various boosters of the dam (Phil Swing and his congressional allies, the Bureau of Reclamation personnel, local boosters in the southwest) is all characterized by what I call a modernist imperialist discourse that emphasized the role of the state in building the dam. These groups discursively constructed the river as a threatening and unpredictable stream whose powers were flowing to waste. In the spirit of Progressive-Era programs, the river needed to be remade by the federal government into a natural resource for all society through the rational application of science and technology. Opponents of the dam were also imperialistic and modernist in their rhetoric, and supported the construction of some sort of dam, but contested the strong role of the federal government. I thus term them defensive-modernists. Reacting to government proposals to distribute the power generated at Boulder, opponents suggested instead that private power companies should reap the benefits of the river's force.

Ultimately it was the state-sponsored plan that won out at Boulder:[1] the federal government would put forward the money and the design, private capital would contract to build the dam, power would be leased to private utility companies for distribution, and throughout several strikes and work actions the job site would remain non-unionized. This chapter will explore these battles, showing how there was little opposition to the building of a dam at Boulder Canyon. Instead, discourse centered on what *form* the dam would take—in effect the arguments were not over *whether* humans would dominate nature, but the *manner* in which domination would occur.

Boulder Dam inaugurated a golden age of dam building in the United States, dated loosely from 1930 through 1960. In his journalistic style, Marc Reisner (1986) called this time "the Go-Go years" while the more academic Lawrence Lee (1980) calls it the "Second Phase of Reclamation." Boulder Dam started this period as the first high dam proposed and built explicitly for multiple purposes. The legal and technical groundwork established with Boulder Dam determined the course of the other large dams; similarly, the social and political maneuvering required to construct Boulder Dam informed the discourses around Grand Coulee and Glen Canyon.

The golden age of dam building was based on a revolution in technology and on changes in the socio-political environment. Nineteenth century dams were generally small projects that had local effects. Dams from Maine to Minnesota hold back small rivers for local sawmills or other diminutive irrigation or power-production projects. They are generally low dams, modeled on the natural example of the beaver dam. They are 5 to 50 feet high and in most cases have only a single, generally economic, purpose.[2] Dams in the West, on the other hand, hold back larger rivers that often flow through deep canyons. As a result, dams as tall as 300 feet are sometimes characterized as "low." With improvements in engineering technology, the height of the dams increased, and a qualitative change in labeling became accepted. When the 726 foot-high Boulder Dam was built—the highest dam in the world at the time—it inaugurated the era of the "high" dam.

AN INTRODUCTION TO THE SOCIAL, POLITICAL, AND GEOGRAPHIC HISTORY OF BOULDER DAM

Boulder Dam was preceded by a fifty-year history of dam building and federal programs that attempted to control nature, although on a much smaller scale. The federal government, individual states, and private capitalists had all attempted to harness the power of rivers, turn the water to irrigation, and alter streams to make them navigable. This section will set the context for a more in-depth discussion of the discourses about Boulder Dam.

In his 1901 State of the Union address, Theodore Roosevelt pointed out that private irrigation measures had largely failed—the federal government alone was capable of coordinating the large projects that the geography of the West demanded. The federal government also had title to much of the land that projects would be built on. Boulder Dam was one of the first projects that the Bureau considered after the Reclamation Act of 1902, and it was certainly one of the grandest in scale. Unfortunately for the new Bureau, the hurdles to cross before the dam could be constructed were as grand as the dam itself.

The Boulder Canyon Project endeavored to tap the Colorado River to bring water to the parched Southwest, and intense interstate bargaining preceded construction of the dam. Because it is a huge, multipurpose dam that sits on a river flowing through several states, the Bureau had to navigate through the morass of interstate politics and long time western rivalries. The solution to many years of interstate squabbling was found in the Colorado River Compact of 1922.

The Compact was a deal that negotiated where water would be delivered in the various states, and was first signed by six of the seven states involved: California, Colorado, Nevada, New Mexico, Utah, and Wyoming, with Arizona dissenting (Hundley 1975). Arizona attempted to stonewall the project because of perceived inequality in the allocation of the water. Following Theodore Roosevelt's restatement of the Utilitarian logic, the greatest need for the water was in California—mostly Los Angeles—and so a disproportionate amount of the water was given to California. Los Angeles was also given any excess water that went unused by the upstream states. But soon after the signing by the six other states, in 1928, Congress passed legislation stipulating that the project could go forward with approval from just six of seven states. Arizona signed on when it became clear that it either accepted the terms or got left out entirely. The deal was brokered by some of the most powerful politicians of the time, including future president Herbert Hoover, for whom the dam would be named.[3]

The Compact represented several significant changes in the federal presence in the West (Abbott 1994). First, the Compact showed what a successful federal intervention could do—it provided an unprecedented compromise between states on the complicated issue of water law. Second, the Compact initiated a debate over whether hydroelectric power was more appropriately administered by private or public agencies, and if public agencies, then whether federal, state, or local government would be responsible for regulation. Therefore, the Compact made possible a series of large dams, starting with Boulder.

Boulder was the first dam to take advantage of new technology that enabled large multi-purpose dams. It provides flood control, drinking and irrigation water, and hydroelectricity. Boulder is truly an engineering marvel. It is 726 feet and five inches high, but it took only five years to build (1931–35)[4] in the middle of the Great Depression, completed two years ahead of schedule and officially 14 million dollars under budget. The construction was handled by a coalition of six different construction companies, imaginatively named Six Companies. This group came in with the low bid of 49 million dollars. By the time Boulder was completed its monetary cost[5] would total almost 57 million due to changes in design, between 20 and 35 percent of this was profit for the Six Companies (Stevens 1988: 252). The two main leaders of the consortium called Six Companies, Henry J. Kaiser and Warren A. Bechtel, went on to build some of the largest construction projects in the West, and later the world (McCartney 1988; Stevens 1988).

Boulder Dam laid the social, legal, and technological groundwork for the other, higher dams that followed, but Boulder still stands as a preeminent symbol of power and control—what David Nye (1994) has called the technological sublime. It is a tourist attraction that people visit from all over America and the world as they see the sights of the West. Its immense size speaks to its prominence as a symbol of the domination of nature—a highway runs over its top, and its base in bedrock is three city blocks thick. The dam dominates the river: stops it cold. It turns the river to ends that society dictates: crops, power, drinking water, recreation.

This value orientation—in which nature is subordinated to society's needs—was played out many times in the West. Shasta Dam on the Sacramento River in northern California, and Bonneville and Grand Coulee Dams on the Columbia River in the Pacific Northwest were under construction as Boulder was completed, and many of the workers simply moved north to continue their newly-invented trade as dam builders. But it was Boulder that launched them all.

THE DISCOURSE ON NATURE AT BOULDER DAM

In order to build a dam—to put a concrete plug in a river and thereby dramatically change the upstream and downstream ecosystem, altering the navigability and the aesthetics of the river—interested parties must justify the changes. This may be more or less explicit given the variability of opposition to the proposed structure, but in all cases there is at least some legitimation of the expense, if not the changes wrought by the dam. In the case of Boulder Dam, justifications were made in terms of the improvements that would accrue to society.

These benefits were compared to the costs of building the dam according to a strict economic logic. There was little consideration of any aesthetic or natural value, for the preservationist arguments had recently been defeated at Hetch Hetchy; instead, the dam was legitimated by devaluing what existed at the time that the dam was proposed. Perhaps most striking was the way in which boosters of the dam discursively constructed nature. This discourse about the river echoed the ways that Americans had come to talk about their forests, their underground minerals, and even their own labor. Such resources were to be used for the greatest good: they were to be scientifically and rationally regulated. In the case of a national resource such as the Colorado River there were many different discourses on the proposal and building of the dam, but all of them rested on a singular and monolithic discourse on nature.

This monolithic discourse was, however, crosscut in certain ways by those who discussed building a dam. The river was primarily characterized as violent, but also seen as providential in certain ways. The dam, proponents argued, was necessary to maximize the productive forces of the river, and to minimize the destructive powers. In short, as Bureau of Reclamation commissioner Arthur Powell Davis claimed, Boulder Dam would "convert a natural menace into a natural asset" (Davis 1924). This discourse on nature is based on an understanding of economic value as the only real measure of worth. There is no appreciation of other values that might be lost in the transformation of the river into a workhorse for society.

This discourse infused with an economic logic was not the only one in circulation at the time that Boulder Dam was proposed. John Muir and a small band of allies had fought against a proposed dam at Hetch Hetchy Valley by the city of San Francisco on the grounds that a dam would destroy a scenic wilderness. That wilderness, furthermore, was supposedly protected under national law through the Valley's inclusion in Yosemite National Park. Muir and The Sierra Club proved unsuccessful in the bid to stop a dam at Hetch Hetchy, and by the time Boulder was being debated, this discourse had lost legitimacy. In some ways, Muir's loss at Hetch Hetchy enabled the discourse on nature used at Boulder to become dominant. Since the opposition had been vanquished, the ways in which Muir and his allies constructed nature—as an aesthetic resource not as an economic resource—was also vanquished. Furthermore, Muir died shortly after the Hetch Hetchy battle, and with his passing the nascent wilderness movement lost an energetic activist. Without Muir the several other core activists were, perhaps, less able or less willing to manage the intricacies of national politics (Clements 1979; Nash 1967/1982; Richardson 1959). This preservationist movement faded with the rise of the conservationist movement in the 1920s for complex reasons, in part due to the strong presence of conservationists in powerful governmental positions. The preservationist perspective, however, will resurface in the 1950s (as discussed in chapter six), but in the late 1920s, at Boulder, the domination of nature and the turning of the river to society's benefit were uncontested.

In 1935 George Pettitt, a San Francisco lawyer, wrote a book entitled *So Boulder Dam Was Built,* published by Six Companies. The author declares that the reason for building Boulder Dam was to rescue the Imperial Valley from flooding:

> Ages ago a river lifted itself into the air by its bootstraps and the sea into which its turgid waters formerly swept dried up and became a desert. Human beings found this ancient seabed, and recognizing the potential fertility of its sun-baked silt, they marked it for their own. Millions of dollars and decades of labor were spent in converting its barren wastes into one of the garden spots of the world. Five hundred thousand acres of land were brought under cultivation and made to yield crops such as even the omniparous[6] Valley of the Nile could not duplicate. Roads were built, cities grew, and seventy-five thousand people came there to secure a livelihood. But always the river that had lifted itself into the air roared and grumbled over head. With every flood it threatened to burst its lofty channel, reclaim its abandoned territory, and in one great deluge destroy all that Man had wrought. So Boulder Dam was built (Pettitt 1935: 6).

Pettitt's book exemplifies the discourse on nature that was common at Boulder Dam. The river is simultaneously a powerful actor (it lifts itself up into the air) and a benevolent provider (its potential fertility and bountiful and varied crops). Humans build upon this landscape and bring forth a bounty that has never been seen, surpassing the great ancient civilizations. Yet the river also continuously threatens this wealth. Nature, in this almost Biblical picture, is both the great provider and the great destroyer. Pettitt's work exemplifies the complexity of the discourse on nature around Boulder Dam. Pettitt portrays nature in a similar way as Machiavelli's *fortuna,* as discussed in chapter two, where *fortuna* is associated with both a providential Christian God and also the Egyptian goddess of destruction, Isis. The domination of nature is not done out of purely utilitarian logic, but is complexly related to economics, politics, and religion. Pettitt can argue that the river is at once the great destroyer and the great provider, and this allows him to take pride in subduing the river and turning its destructive capacity to the ends of human society, enhancing the productive qualities. In this way the alteration of the river could be construed as a fulfillment of the Christian God's benevolence.[7]

Pettitt illustrates his work using several photographs with creative captions that expose this domination of nature. One photo of the dam site with dynamite in the foreground (see Figure 4.1) is captioned "A little powder changes the face of Nature miraculously." Another photo (Figure 4.2) of a dynamite blast is captioned "The Colorado is given a lesson in efficient rock carving."

Because Pettitt is interested in boosting the accomplishments of the dam and its builders, he emphasizes the role of nature as a powerful actor. Nature, to Pettitt, is an agent that needs to be subdued. But in Black Canyon, nature is wild and powerful and does not knuckle under easily. His third chapter is entitled "Rivers Don't Die Without a Fight." Of course only an active agent is capable of death:

> Although rivers often appear consciously delighted in opposing the will of Man, there is no dependable evidence that they observe, think and act with the purposefulness of sentient things. But if they do, as those who work on them believe they do, the mind of the Colorado must have presented a most interesting subject for a psychological study during the first weeks of May, 1931 (Pettitt 1935: 53).

A little powder changes the face of Nature miraculously.

Figure 4.1 Changing the face of Nature (Pettitt 1935).

Figure 4.2 Lessons in rock carving (Pettitt 1935)

For a commentator to describe a river as appearing conscious may seem remarkable from the perspective of a reader at the beginning of the twenty-first century. Nature is commonly thought to be active in only a non-sentient way. In the dominant contemporary worldview, nature is non-purposive in its action: though some may believe in creationism, to most people in the modern West, natural laws animate the world, not spirits. Rivers, in short, are not conscious. They do not take delight, nor do they actively oppose. Pettitt, in the quote above, nods to these present-day understandings when he points out that there is no logical evidence that rivers engage in purposive action. He then, however, claims that the experience of those who work closely with rivers has led them to believe in a river's consciousness. Armed with this legitimacy, Pettitt dramatically suggests that the Colorado is in need of psychological study. Perhaps this is fitting seeing as how Pettitt views the work being done as putting the river to death!

Pettitt's work has many complex layers to it. He makes the case for the dam on the basis of its social and economic gains, yet his book was published by the same company that built the dam, so there is a clear bias to his argument. Nonetheless, Pettitt's work helps to discursively construct the river in a way that justifies the dam and legitimates subduing the river. The complexities of Pettitt's work, then, lay in the relations between utilitarian subjugation of nature simply for society's ends, the interests of the state in gaining further control over the land and securing a productive economy, and the individual interests of the construction company that published Pettitt's book. Pettitt was not simply a utilitarian, and not just a tool of the state or private capital. He was a San Francisco lawyer and soon to be university administrator who had little obvious economic interest in the dam itself. Yet he is an example of the dominant perspective of the time regarding society's relationship with nature. It is a deep and complex relationship, but one defined by domination.

Newspapers also took up the hue and cry of subduing nature. Journals from San Francisco to New York proclaimed the river "harsh," "mighty," and "difficult." Regarding the first proposals for Boulder Dam, on 7 March 1921, the San Bernardino *Telegraph* wrote "Colorado To Be Shackled, Turn Desert Into Farms." In a series on the building of Boulder Dam, the San Francisco *News* claimed on 1 February 1932: "Hoover Dam Builders Thrive on Tremendous Difficulties to Subdue Mighty Colorado," and the next day, "Engineering Magic Conquers Nature's Harshness." The *News's* competitor, the *Chronicle,* proclaimed "Man's Skill in Conquering Mighty River" on 24 November 1931. In October of that year the New York *Evening Journal* ran an article entitled "Gigantic Boulder Dam Task Provides Thrills on Cross Country Auto Tour." The article describes the job site in terms such as "dizzying," "full of excitement," and the Colorado itself as a "fighting and wild, crazy river." *Life Magazine,* on 23 October 1944 devoted a pictorial spread of 13 pages to the Colorado River, entitled "The Colorado: A Wild and Beautiful River Is Put to Work for Man." While portraying the river as wild and destructive, the newspaper headlines nearly cry out that the Colorado must be controlled.

The *New York Times* even printed a poem by William Tell Fogle, entitled "Colorado You'll Be Dammed:"

> Million acres patiently waiting,
> For your waters that are wasting;
> They to feed the many peoples
> Crowding toward the Western ocean,
> Where the glowing golden sunshine
> Makes life worth the living;
> Free from cyclones, snow and blizzards,
> Frost-bit feet and weather wizards:
> Oh! Colorado, you'll be dammed.
>
> Then up rose Swing and Johnson,
> Champions of our loved Southland.
> They among the nation's solons,
> Stood like mighty, battling Trojans,

Speaking, fighting for their bill
Which would curb the treacherous waters,
That menaced the trusting farmers
Of Imperial Valley's fertile vale:
Ah! Colorado, you'll be dammed.

This poem, only a portion of which is quoted above, simultaneously expands on the ways in which nature is destructive (treacherous waters menace the trusting farmers) while recreating the myth of southern California as a climatic paradise. The fertile farmland is waiting for the champions of the southland to rise up and save the wasted waters. The champions are Swing and Johnson, authors of the Boulder Canyon bill, and they are the heroes of the trusting farmers who are so threatened by the treacherous waters of the river in flood. This poem overflows with hyperbole, characteristic of much of the discourse about Boulder Dam.

Newspapers also portrayed the natural area that the reservoir would inundate as a waste land. As the San Francisco *Daily News* reported on 19 February 1924: "Boulder Site Desert Waste: Aged Prospector and Wife Only Inhabitants at Present Time." The article opens with the image of a rugged and harsh area, countered with the grandeur of the project: "An aged prospector and his wife, living in a tent on the river bank are the only inhabitants of the imposing spot which some day will be the scene of the greatest engineering project in man's history." Such rhetoric helps to construct the area as unworthy of anything except this history-making project.

Representations of the river as harsh or temperamental were common not just to the popular press or industry boosters. The Bureau of Reclamation, Congress, and the Office of the President also engaged in such reconstructions of the river as wild and untamed, yet able to be converted to human purposes. In reports, inter-office memorandums, and presentations to Congress, the Bureau of Reclamation continually portrayed the river as, in the words of Commissioner of Reclamation A. P. Davis, "a natural menace" that could be converted into a "natural asset" (Davis 1924).

A similar logic and rhetoric infuses the report written by a select committee called the "Special Advisors to the Secretary of the Interior," which was asked by Secretary of Interior Hubert Work to review the dam proposals in 1927. Their report, filed in 1928, suggested that the dam was most feasible at Black Canyon, and that:

> the Colorado River is a natural resource of great potential value by reason of a combination of abundant water supply with characteristics of drainage basin and stream which presents striking possibilities for the development of agricultural lands through irrigation, the generation of

> power and the furnishing of water for important municipal, industrial, and other uses. The need is now apparent for a major step forward in the development of the Colorado River and in its transformation from an instrumentality conveying grave menace of destruction of life and property to one of much greater usefulness than now effected. (Special Advisors to the Secretary of the Interior 1928: 366).

It is apparent in this quote that the Special Advisors (a group that included several governors of western states, university professors, and a former Secretary of Interior) viewed the river as a tool for the use of humans rather than as something with inherent scenic or aesthetic value. The river is a "destructive" "menace" which can be put to more effective and efficient use through governmental intervention. Viewing the river as an instrument that contains value only so far as human economic purpose allows the extension of a high modernist project to "transform" the river to "great usefulness."

Ed Sample was a State Senator from San Diego who he ran for Congress in 1924. He took up the development of the Colorado as part of his election platform, declaring (as quoted in an undated and uncited newspaper clipping in the Phil Swing Papers at UCLA[8]):

> The people of the nation have become alert to the fact that the Colorado River is one of its greatest resources. Its fructifying waters, when spread upon our arid acres, will produce food for hundreds of thousands. The restless energy of its cascades will light and warm a million homes and turn the wheels of an empire's industry. To develop and use the nation's fuel is to impoverish; but to develop and use Colorado's power is to establish contact with the Infinite and avail ourselves of that which enriches without being consumed, and is renewed forever. Accordingly, the problem of the Colorado's development calls for the application of the highest statesmanship and for a vision which comprehends the nation. To conceive of it as special or a local or partisan is to be unworthy of the splendid gift.

Sample's rhetoric is full of superlatives, but it is based on the idea that nature must be dominated so that the wheels of an empire's industry can be turned. This is the rhetoric of a modernist—one who believes in the transformation of nature to benefit humans, so that society may progress—but it is also the rhetoric of one who believes in a higher power, "the infinite." Jackson Lears (1981) has shown how the complexities of the transition to modernism can be expressed by a backwards-looking search for meaning, or a reliance on mysticism. The transition to modernism, Lears argues, is crosscut with antimodern tendencies, as seen in Sample's mysticism.

Given the lack of rhetoric regarding any inherent value of nature, it is interesting to note Sample's construction of the issue around the idea of conservation. To Sample, the power produced by the river is renewable compared with the development of fossil fuels. Indeed, those who would dominate the river and turn its power to the ends of society often used the idea of conservation as a justification. As discussed in chapter three, this idea of conservation is at odds with ideas that were contemporaneously characterized as preservationist. Conservation, in this context, implies society's use of nature in the utilitarian mode of maximizing benefits for as many people as possible. For Sample this means hydropower over the use of fossil fuels.

For others at the time, including Franklin D. Roosevelt and Harold L. Ickes, similar ideas were contained in the concept of cooperation between society and nature. The idea of cooperation, however, highlights a certain respect for nature rather than outright domination. This is evocative of Marx's concept of the cooperation between labor and capital: they cooperate within the process of production, "with capital supplying the physical technology and labor the human energy" (Schnaiberg 1980: 210).

Harold L. Ickes, Secretary of the Interior for Franklin D. Roosevelt, was a tireless promoter of the ideas of conservation, and understood the special place that Boulder Dam had in changing the relationship between humans and their environment. In his speech at the dedication of Boulder Dam on 30 September 1935 (during which Ickes took great pains to drive home the official name of the dam), Ickes pointed out that never before had such a structure been built for the economic purposes of humans. It was, for Ickes, an example of how humans and nature should relate:

> Through bitter experience we are learning the lesson that it is not through the destruction of nature or even though its complete subjugation that we will be able to serve our own greatest good; that if we would avoid the tragic mistakes of the past and assure for ourselves and for the generations that are to follow us the material and spiritual benefits which will flow from a wise and prudent use of the rich resources of nature, our approach must be in a spirit of cooperation. . . . No better example of understanding cooperation between man and nature can be found anywhere than this imagination-stirring project that, in grandeur of conception and in skill and speed of execution, ranks as one of the greatest engineering undertakings in the history of the world. And what more characteristic and appropriate name could be chosen for this monumental enterprise than the one with which it has been christened? As the eye encompasses the majesty of this work and comprehends the bold and rugged setting chosen for the taming of the waters of the turbulent Colorado, the mind appreciates that this setting and this accomplishment of what at first must have been regarded as the conception of

> a fevered imagination would not be worthily and fittingly named by any less bold and striking designation than that of Boulder Dam.[9]

Ickes characterizes the Colorado as "turbulent" and "wild" in a "rugged" setting, an appreciation of which must lead to the approval of the name "Boulder Dam." This setting stands in contrast to the dam, which is "majestic" in its "grandeur." Ickes portrays the benevolence of the government in transforming the rugged setting and taming the raging waters of the river. Nature, to Ickes, is a resource among many other resources, and one that should be subjected to the rigorous logic of utilitarianism as with so many other aspects of efficient government. In short, the river should be used in such a way as to maximize the greatest good for the greatest number:

> I venture to hope that this dam, with its great storage of health and wealth and happiness for thousands of people, will stand as the definite opening of a new era with respect to the natural resources of America—an era of conservation which means the prudent use of all our natural resources for the greatest good of the greatest number of our people; an era that will recognize the principle that the riches of forest and mine and water were not bestowed by God to be ruthlessly exploited in order to enhance the wealth of a small group of rugged individualists, but were beneficently given to us as an endowment to be carefully used for the benefit of all the people. On no other theory would the Federal Government be justified in so generously opening the doors of its treasure house for the building of this and other similar projects that will turn large sections of this breathtaking western country into rich homesteads where a happy and contented people will find it possible to live those comfortable and worth-while lives that we covet for every man, woman and child in these United States.

Discourses on nature at Boulder Dam deconstructed the river and reconstructed it as a resource for use by society. Ickes' statement about the "cooperation" between nature and society underscores the relevance of the specific case of the Colorado River to other cases of natural resources—forests, minerals, or animals—as well as the general idea of nature. Actors at the time saw, in the Colorado, a "menace" and a "tyrant," but they also saw an opportunity. The Colorado, as with Ickes' words above, was a natural resource given freely to society, similar to America's forests, gold, the bison, or the beaver. Humans were to cooperate with nature in this opportunity, and the only limitation put on rational and democratic societies was that the benefits be maximized for the greatest number of beneficiaries.

Such rhetorical justifications for building the dam expose an understanding of the relationship between society and nature as one of exploitation and use for human benefit. As discussed in chapter two, the gendered character of much of this discourse is striking. The construction of the river as a "natural menace" that could be transformed into a "natural resource" was a crucial development for the continuance of state building in the American West, as will be seen in the next section. This reconstruction was accomplished through a complex set of discursive negotiations among boosters and detractors of the dam. Local farmers, urban populations and their municipal administrators, and many Progressive politicians supported the dam on the basis of its benefits to be granted to the public. Opponents decried the public intrusion into the private utility industry, but remained in favor of building the dam as a way of furthering development and "progress." Nowhere to be found was a discourse that valued the river *qua* river, as an entity with aesthetic value unto itself.

BOOSTING BOULDER: THE DISCOURSE OF IMPERIAL HIGH MODERNIST STATE BUILDING

To make sense of the discourse on boosting Boulder Dam, we must understand the political situation in the Southwest. The impetus for Boulder Dam came primarily from engineers in the Bureau of Reclamation and farmers in the Imperial Valley of California, just across the Colorado River from Arizona on the Mexican border. This fertile valley produced many forage crops, fruits, and vegetables. It also produced a class of farmer that was willing to call on the federal government for help in building irrigation systems. Thus, Boulder Dam started its political life as a small local project in southeastern California along the Colorado River.

Imperial Valley, the All American Canal, and the original boosters

The area now known as the Imperial Valley of California, the very southeastern corner of California, was originally called the Colorado Desert. Oliver Wozencraft noted the valley as potential agricultural land in 1850, and formed a development corporation with Charles Rockwood, an irrigation engineer. Wozencraft and Rockwood attempted for many years to gain title to the public land, and then to build an irrigation system, but it required the additional expertise of George Chaffey to develop a viable water system. Water came to the valley in 1901, and Chaffey re-christened the area the Imperial Valley to give it a more attractive name for prospective emigrants (Alexander 1928; Holt 1907; Holt 1908; Kennan 1917; Rockwood 1918; Schonfeld 1969).

By 1904, demand surged as more and more land came under cultivation. Due to silting up of the gates and ditches, the amount of water delivered to the settlers could not keep pace (Rockwood 1918). Rockwood had to increase the flows because legal claims could be made upon his company should the water levels drop so low that irrigation could not continue. Rockwood was a great booster, if not a bit of a huckster, but he was not a great engineer. Between a rock and a hard place, Rockwood made a new diversion cut in Mexico, and in doing so ensured the demise of his company. In the few years since Anglos had settled the area, rainfall had been slight and fairly regular, but in 1905 the rain came, and it didn't stop. The Colorado rose up the levees and diversion canals, until it gouged away the dirt mounds and broke the banks. Within a few days, the full force of the river flowed through the cut Rockwood had just made and into the largest and deepest sub-sea level area in the hemisphere—the area formerly known as the Salton Sink averages 200 feet below sea level. The full flow of the Colorado ran straight downhill into the Salton Sink, forming what is now called the Salton Sea.

The Colorado Development Company worked to repair the breach of the Colorado, but it took three years before the river was made to flow in its historic channel. The full flow of the Colorado is about 15 million acre-feet of water per year. For three years this water flowed into the Salton Sink—almost enough water to cover the land area of the state of Wisconsin at one foot deep. Rockwood was forced to surrender his ownership of the water systems and the water rights, leading Southern Pacific Railroad (SPRR) to step in. The massive engineering power and wealth of the railroad was finally successful in closing the breach—but only after almost three years of work (Kennan 1917; Schonfeld 1969). The SPRR continued to manage the irrigation system for two years, and in this time growers repaired the damage to fields and started growing crops again. In the course of the flood, deep furrows—small river channels, really—were dug out of the valley floor. These were turned into drainage ditches for agricultural runoff, enabling irrigated agriculture to expand even in the face of such clear potential for disaster (Dowd 1951). These flood-channels-turned-drainage-ditches are now called the New River and the Alamo River, giving the Salton Sea two inlets and no outlets. Even in the presence of these threats, pioneers came to the Valley. Between 1901 when the first water flowed in, and 1910, the population grew to 13,591 (Rudy, 1995). These eager farmers were rough-hewn men. Most lived on their claimed land in shanties or tents, many of them barely had the capital to buy their own farming tools (Stuart 1961).

In 1906, very soon after water started flowing into the valley in a controlled manner, relations with Mexico grew tense as the Mexican revolution approached. This led growers to realize the need for an effective political lobby whereby they might protect their interests. The farmers found this muscle in mid-nineteenth century California water law, and in 1911 they voted to form an Irrigation District. Irrigation Districts are quasi-private: members must own land in the District boundaries. Each landowner gets a vote in the decisions of the Irrigation District, making the organization formally democratic. This type of democracy, however, is certainly not progressive. It is more similar to Ancient Greek democracy than modern liberal democracy: each landowner has a say in how farm land is used, where irrigation water is to be sent, and other decisions that are vital to the economic and social health of all the citizens of the entire county. The irrigation district organization clearly excludes those workers and citizens who do not own property. Donald Worster (1985) uses this example of pseudo-democracy to point out the domination wielded by elites and businessmen who controlled the water and the capital in the Valley. Worster points out that few small farmers held seats on the board of the Imperial Irrigation District (IID), and he describes a division between the "insurgent mass of farmers" and the businessmen who were "intent on using their position as water dispensers to keep the right people running valley affairs" (Worster 1985: 201). While it is true that the first years saw many bankers and businessmen on the board of the IID, by many accounts it appears that the culture of the Valley was such that lawlessness prevailed in the face of organized elite domination (Stuart 1961; Woodbury 1941).

During the first years of the IID there was often unrest and raucous goings-on (Stuart 1961; Woodbury 1941). Meetings of the IID often took the form of a circus, and starting in 1911 a farmer by the name of Mark Rose began making regular appearances to lobby for his idea for Imperial Valley development. Rose owned land on the east mesa of the Valley—land that was too high above the present irrigation systems to receive water. Since he was a landowner and a member of the association, Rose believed that he was due both water and his share of the wealth being generated. Rose's idea was to make a new diversion from the Colorado, north of the Mexican border. This diversion would feed a newly constructed canal entirely on U.S. soil, solving the troubles involved with the present canals that flowed through Mexico. Such a canal would also be next to Rose's land, thus providing him with water (Woodbury 1941). Rose was routinely laughed out of the meeting hall for such an expensive and selfish scheme. It seemed clear to the association members that it was primarily Mark Rose who would benefit from this idea.

Rose kept hammering on the association about his idea until finally in 1917 he saw his chance for a coup. Rose came to the meeting with a few supporters, and sat through the usual dismissal of his idea by the board. At his signal, Rose's supporters went up to the stage with ropes, and Rose shouted that any board member who did not resign immediately would be lynched. They all resigned, and a new board was created on the spot—with Rose as president (Woodbury 1941). Rose started work on his new project—now dubbed the All American Canal—and buttressed by the continued rumblings of revolution in Mexico, he soon had many supporters. Since water went through Mexico to get to the Imperial Valley, it was subject to the demands of corrupt Mexican officials.

One of these supporters was Phil Swing, the legal defense for the IID. Swing was convinced about the canal's importance, and began to work actively with Rose on the project. In a turnabout from several years before when they felt betrayed by the Bureau of Reclamation's attempts to take over their project, in 1919 Rose and Swing invited the U.S. Bureau of Reclamation to make a study of the project (Bureau of Reclamation 1922; Bureau of Reclamation All American Canal Board 1919).

Arthur Powell Davis was a sympathetic ear in the Bureau of Reclamation. Son of John Wesley Powell, the famous one-armed Civil War Captain and Geologist who first mapped the Colorado River in the late 1860s, Davis had different plans for the river than rafting and mapping it. He envisioned a grand series of multi-purpose dams holding reservoirs of agricultural and drinking water for seven states. A troubled Bureau of Reclamation needed a successful new project; the Bureau's few dams and canals from its first 15 years of operations had encountered problems. The beneficiaries of the water projects had defaulted upon loans, and unfriendly members of Congress were looking for a chance to gut the Bureau's funding (Reisner 1986). The growers of the Imperial Valley just wanted a canal to bring them water that didn't have to travel through Mexico. They got much more than they bargained for.

Swing and Rose realized that they would need an advocate to get their All American Canal through Congress. They traveled to Washington to lobby for the canal, with Swing writing several versions of a bill to procure funds for the Bureau of Reclamation to build the canal. Each one, in turn, died in committee (Woodbury 1941). After several years of trying, Swing realized that the farming interests of the South and Midwest were against him. Any new lands opened up in California meant more competition for these primarily agricultural states. In 1919 the Imperial Valley farmers decided to try a different tack. Their new idea was to send one of their own to represent them in Washington and push their bill though. This lucky person turned out to be Phil Swing. He was elected—by a landslide—to the U.S. House of Representatives in 1920.

Boulder Dam becomes the focus of discourse

Swing's election marked the beginning of official discussion of Boulder Dam itself, as well as a shift to discussing the Colorado River as a resource for multiple uses. Representative Swing realized that the All American Canal was too small a project and without broad appeal. It became clear to him that in order to push through the canal, it would mean developing the entire lower Colorado River (Dowd 1956). A small project that benefited local boosters was transformed into a national project that was said to benefit all Americans. With this shift, the Imperial Valley interests came into line with the Bureau of Reclamation and most of the Western states. Swing rewrote his bill with the senior Senator from California Hiram Johnson, and it was finally passed in 1928. During this eight-year period between Swing's election and the final passage of the bill, the discourses on Boulder Dam, both local and national, became intense.

The discourses used in this long and arduous journey from small-scale canal proposal to massive multiple purpose river development project were many and varied, but can be classified into several categories—flood control, irrigation, and hydropower—the common element of which, as previously discussed, was to construct the river as an entity to be dominated. The boosters of the project continually constructed the river as a menace—an unpredictable and wild river that was capable of flooding land and destroying businesses and towns. The flood control argument was the most overtly hostile to the river as it was said to pose a threat to life and livelihood. When boosters used these arguments they invariably characterized their goals as noble and righteous: to subdue the river and turn it into a predictable and tame stream.

Boosters often combined this emphasis on flood control with the productivity that irrigation and reclamation offered. Americans, they argued, could reclaim land that was presently desert. It could be transformed into productive cropland with the addition of regulated water from the Colorado River. This water, presently, was flowing "wasted" to the sea.

A further benefit to be gained from a high dam on the lower Colorado was hydroelectric power. The technology for hydropower was just being perfected in the 1920s, and boosters argued that this additional purpose of the dam would so greatly benefit the Southwest that the federal government should build the dam at taxpayer expense, to be paid back with the sale of hydropower. These several key uses were collected together for the first time in the recommendations by Albert Fall and Arthur Powell Davis in their 1922 report to the Department of Interior, known as the Fall-Davis Report (Bureau of Reclamation 1922: 21):

> It is recommended that every development hereafter authorized to be undertaken on the Colorado River by Federal Government or otherwise be required in both construction and operation to give priority of right and use: First. To river regulation and flood control. Second. To use of storage water for irrigation. Third. To development of power.

On the basis of these discourses, boosters constructed the Colorado River as a resource for multiple uses. These discursive constructions of the river as a potential resource that needed taming allowed the state to build an infrastructure on the landscape—to dominate nature to an extent never seen before. Though there are many different strands in this discourse—flood control, irrigation, hydropower—they all converge on the central idea of state-sponsored infrastructural development. The federal government should build Boulder Dam, the argument went, so as to extend the benefits of nature to society, to develop the economy of the West, and to provide important services to the population. In the language of state building, this means a combined domination: the state makes efforts to control nature, control the economy, and control the population. The booster's own words best characterize this state-building discourse. In controlling floods, reclaiming land through irrigation, and producing hydropower, the boosters argued that they were building an empire in the West. The best label for this discourse is thus *imperialistic*. The several strands of this imperialistic discourse, woven together as strands are woven to form a cable, all contribute to the discursive construction of the river as an economic resource that will promote economic and social progress in the West. The three main strands of flood control, irrigation, and hydropower were at times joined by an argument that the benefits should be given to Americans only (the All American Canal would avoid delivering water to Mexico), and that those Americans who were most deserving were ex-servicemen. These elements were drawn together to form a generalized argument of "national importance" that highlights the imperialism of the discourse.

Flood control

Flood control was perhaps the most useful discursive tool for the boosters of Boulder Dam. Though it was one part of a larger economic cost-benefit argument to justify the construction of the project, controlling floods had a strong human element to it: the brave pioneer was threatened through no fault of his or her own. The wild and unpredictable river menaced them, but through the Boulder Canyon Project this could be stopped. The boosters completely ignored that this "natural disaster" of flooding was entirely produced by human actions (building communities in a flood plain, making bad engineering decisions).

Commissioner of Reclamation A. P. Davis wrote a persuasive letter to John Bacon, member of the California Colorado River Commission and Mayor of San Diego on 3 June 1922, outlining his reasons for "the urgency of the construction of the Boulder Canyon reservoir." The letter followed a severe flood in the Palo Verde Valley of southeastern California, and first among Davis' many reasons was flood control. Davis called the flooding of the Colorado River "annual and imminent," and noted the high costs paid by the people of the Imperial and Palo Verde Valleys to fight these floods. This flood "menace," Davis argued, will continue to grow, threatening the settlement and use of the valleys.

> If the Colorado should flow uncontrolled into the Imperial Valley until it was completely inundated it would destroy the valley entirely for any use that could be made of it by the present generation or by their children or grandchildren. . . . The construction of the Boulder Canyon reservoir would so reduce the flood flow of the river that it would no longer be a menace to these valleys.

Davis clearly argues that the control of the Colorado is necessary for the protection of the Imperial Valley. Lacking in this, of course, is any questioning of why this valley was settled if it was under continuing and imminent threat of destruction.

The "menace" that the Colorado posed to the farmers of the Imperial Valley made for great press coverage. News of natural disaster helped prove the "menace" portion of the boosters' argument. The *Los Angeles Express* reported on a joint committee meeting called by the Governors of California and Arizona on 23 June 1922 to discuss the flooding. The committee reported that

> As a result of our investigations we are convinced that emergencies will continue to occur along the Colorado River between Arizona and California unless a large, comprehensive plan shall be evolved and mutually adopted for the ultimate control of the river within a fixed permanent channel.

Flooding, clearly, is bad for business of all kinds (though perhaps good for selling newspapers). The states had an interest in protecting the population, and presenting an image of safety that was in harmony with the myth of the Southwest as a paradise. What migrants would want to escape a Midwest tornado only to face Western floods? What tourists would want to visit a place of wild, life-threatening rivers?

The Riverside *Press* on 27 May 1922 published a photo of people in a rowboat in the Palo Verde Valley under the headline "Colorado River is Shown on Rampage From Which Some Think it Could be Restrained." The caption below says "Floods like this at Ripley, Cal., could be done away with victims say, if a dam were constructed in the Colorado River at Boulder Canyon." In fact, boosters used this very same picture to help their cause in Congress. (For a similar representation see Figure 4.3, from Kinsey, 1928.)

The magazine *Pathfinder,* after the completion of the Dam, looked back at the undammed Colorado and noted that "through the ages it had never been harnessed and its roaring waters allowed to flow unchecked cut its canyons deeper and deeper. Its periodic floods have caused so much damage that about 30 years ago President Theodore Roosevelt asked Congress to make an extensive survey, the beginning of a project which was completed under his cousin, President Franklin D. Roosevelt" (Staff 1935: 12).

Don Kinsey also referenced Theodore Roosevelt in his book called *The River of Destiny,* published by the Los Angeles Department of Water and Power in 1928. Kinsey put the flood menace in a place of prominence in the reasons for developing the Colorado River, noting that Roosevelt told congress in 1906 of "*the necessity of permanent Colorado River flood control and development work*" (Kinsey 1928: 26, emphasis in the original). Kinsey's work is meant to help boost the dam among the people of Los Angeles, and his rhetoric borders on the superlative: "*The Colorado River is one of the most terrifying forces of wholesale destruction in the United States*" (1928: 19, emphasis in the original). The Colorado, Kinsey argues, is prone to flooding and needs to be controlled if destruction is to be avoided (see Figure 4.3). Kinsey argues that the control of the river is fated: the Colorado is a river of destiny which "*opens wide the door to a new era of prosperity nationwide in its beneficial effect*" (1928: 63, emphasis in the original). The works of boosters such as Kinsey were meant to persuade the public of the importance of developing the Colorado in ways that would improve control of both nature and the economy. Only through developing the "river of destiny," so his argument went, could the nation achieve its "new

FLOOD SCENE IN THE PALO VERDE VALLEY IN CALIFORNIA
WHEN THE COLORADO INUNDATED THIS AREA IN 1922

Figure 4.3 Flood scene in the Palo Verde Valley (Kinsey 1928: 18).

era of prosperity." This new era would also include increases in agricultural production through irrigation. While the issue of flood control played well to the larger crowds of the southwest, it was irrigation that held the money and the political will behind the project.

Irrigation

In tandem with flood control, irrigation measures were a key to the boosting of the Boulder Canyon project. The regulation of the flow of the Colorado, the reduction in silt load, and the construction of the All American Canal would help to bring hundreds of thousands of acres of fertile land into production.

Phil Swing, writing to the Imperial Valley Farm Bureau on 30 September 1921, made this argument quite succinctly: "This is the purpose of the Boulder Dam, to supply water to our lands." Swing's purpose could not be more clear. Swing is, of course, speaking to his main constituency and a group who is interested in the project for only one reason. As a politician, one might expect Swing to reflect that sole purpose back to the group. It is also likely, however, that Swing is being honest here. He was, after all, the lawyer for the Imperial Valley Irrigation District, and the All American Canal was the reason he was elected to Congress. It would appear that the further reasons for development that he propounds to other groups are simply strategies for coalition building.

In such coalition building, Swing's language tended towards the grandiose. The San Bernardino *Sun* ran a headline on 1 December 1921: "Colorado River Project Will Flow Wealth Into Southwest, Swing Says: Irrigation of Millions of Acres of Barren Land Will Compare With Panama Canal for Benefits To American People." Swing portrayed Boulder Dam as a monumental project that compared favorably with the last great national project led by Theodore Roosevelt. In fact, Swing's insinuation is that the Boulder Canyon project may be even better, since the benefits will go solely to Americans.

The Imperial Irrigation District also produced booster pamphlets describing the many benefits to be gained from irrigation in the Valley. In one such pamphlet the IID showed the vast fields with a smiling child displaying the latest strawberry crop (Figure 4.4). The text accompanying the image reads, in part, "There is almost no variety of vegetables or fruit that is not grown profitably here" (Imperial Irrigation District 1926: 26). There is no mention, of course, of the subsidized water required for those profitable crops, or the cheap immigrant labor used from the Bracero Program.

Figure 4.4 Imperial Irrigation District publicity photo, 1926.

The irrigation argument was often propounded alone, but when there were multiple social groups in an audience, other benefits were added. The *Los Angeles Examiner* editorialized on 16 February 1924 that: "If this Boulder Dam is ever built, as proposed in the Swing-Johnson Bill, and as planned by the U.S. Reclamation Service, it will serve two great purposes: furnish water and irrigation to the people of the Imperial and Coachella Valleys, a magnificently rich area equaling several small States of the Union and enable development of 1,000,000 horse-power for the cities of Southern California—power that will cost them far less money than they are now paying to power companies." Indeed, the irrigation arguments and flood control arguments were often combined with the third major use of the project: to produce hydropower for the urban markets of southern California.

Hydropower

The production of power for urban markets was a controversial part of the project, as will be discussed in the next section. Boosters of the project, however, were interested in using the sale of power to help pay the costs of the dam, an innovative new method of financing. This power production aspect caught the imagination of bureaucrats and newspaper reporters alike.

Arthur Powell Davis in his 3 June 1922 letter to John Bacon, Mayor of San Diego, wrote that the dam would:

> produce sufficient power to pay for itself, principal and interest, constitute an excellent investment of its entire cost, greatly stimulate the mining and other industries of the Southwest, and cause a direct savings in the oil supply of 20,000,000 to 30,000,000 barrels of oil annually. The generation of electrical power at the proposed reservoir would have an important relation to the conservation of the oil supply of the nation.

Davis thus viewed the project as a conservation measure. As it would "stimulate" industry and mining, the project contributed to the progress of the nation as a whole, and to the building of an empire in the Southwest. Many newspapers also reported on the issue of boosting power production and the replacement of oil power with hydropower.

The Brawley (California) *News,* for instance, reported on 15 July 1921 that Swing was fighting hard for the project in Congress, quoting Swing as saying "I am using every effort to influence Washington in favor of constructive legislation rather than destructive, and in my estimation the building of battleships at 25 to 30 million dollars each is destructive legislation. I am fighting for the development of the 2,000,000 horsepower in the Grand Canyon now going to waste and here is the slogan I have adopted, 'Drop a battleship into the Grand Canyon and develop the horsepower!'" Swing, in other words, was asking Congress to forego the price of one battleship, for which Boulder Dam could be built. Putting money towards the dam would produce much more than one ship: it would produce power for the war effort.[10] The dam, however, would eventually cost more than twice this amount.

The acting Mayor of Los Angeles, Byle Workman wrote an article for the *Los Angeles Examiner* published on 6 January 1924. He argues that the Swing-Johnson bill is necessary for the development of the area. "Los Angeles, facing a shortage of electrical power now and the necessity for an adequate water supply eight years hence, must get both from the Colorado River floods." Figure 4.5 graphically represents this argument.

The imperialist discourse was thus based on the domination of nature, in this case the Colorado River, so as to improve the economy and control the territory of the West—to build an empire in the Western states for the benefit of all Americans. There were three strands to this imperialist discourse—flood control, irrigation, and hydropower—and when woven together they were powerful tools for boosting Boulder Dam. As will be shown below, the several social groups that organized around the proposal of Boulder Dam all used these three elements in their arguments for the construction of the project. Many newspaper reports combined the three in support of the project.

Figure 4.5 The *Los Angeles Examiner,* 1926.

On 22 December 1928, for instance, the *Los Angeles Examiner* ran a 3 cm high headline reading "BOULDER BILL SIGNED, WORK STARTS AT ONCE." The article read "The big news is this: President Coolidge has signed the Boulder Dam bill, providing for reclamation, flood control, and water power and the use of the Colorado River. One hundred and sixty five millions will be spent as a starter; not much for a country so rich. More important than money is the fact, now established, that the people have the right to use their own water power without asking permission of 'big men in the business' . . . The long, spectacular struggle to have Congress provide the means of transforming the destructive flood waters of the Colorado River into one of the Nation's greatest benefits ended simply." The article included a photo of Coolidge's signature of the bill in front of just five people; hence the ending was "simple." Using anti-corporate language[11] ("the people" versus the "big men in the business") the *Examiner* article aptly sums up the constellation of "benefits" that the "destructive" Colorado River is being "transformed" for: flood control, irrigation, and hydropower. There was a profusion of arguments for the dam—flood control, irrigation, power production—accreted layer by layer into a larger discourse about the building of an empire. The economy of the western states would be improved with the power production, the locals would be saved from catastrophic floods, and irrigation would produce a bounty for distribution throughout the nation. Furthermore, the dam would pay for itself, giving it an "excellent" cost-benefit ratio.

Ultimately, the passage of the Swing-Johnson bill relied on this trio of arguments, but contingency also played an important part: The Teapot Dome scandal was partially responsible for the passage of the Boulder Canyon Project. In the political context of a scandal involving the private appropriation of public lands, public development of national resources gained strong favor, as seen in such national periodicals as *American Globe,* which ran an article entitled "The People of the United States Do Not Want Another Tea Pot Dome Scandal" (Schaefle 1928).[12] Congressman John O'Connor (New York Democrat) opened debate of the final version of the Swing-Johnson bill saying "Gentlemen, when you get through hearing the arguments of the opponents of the measure you will realize . . . that you are again confronted with the question of Government ownership. . . . That bugaboo is always with us, hanging over us, phantom-like—an 'evil spirit' . . . Our government has for the last time placed in private hands any of the great resources of the country. Of all our resources no great natural power development in the country will ever again go into private hands" (quoted in Moeller 1971: 114). A motion was made by a member of the opposition to recommit the bill to the Rules Committee, which lost 219 to 139. On the strength of this tally, Phil Swing's project, introduced four times over six years, finally passed on a simple voice vote.

The context of scandal lent the bill some extra weight for passage, but the discursive construction of the river as a resource lay behind the call for public development. Several important social groups used such discourses. In civil society the most important group was the Boulder Dam Association, but the Association was joined by veterans and many other groups. There was also a great deal of support among members of Congress and the several Executive administrations that witnessed the four versions of Swing's legislation. The Bureau of Reclamation also used many of these same arguments in their internal and public discourse. The next sections closely examine this imperialistic discourse, emphasizing the ways in which the different strands of rhetoric construct the river as an economic resource, whose development will help to build an empire in the southwest.

A vibrant civil society: The Boulder Dam Association, The American Legion, and the discourse of "national importance"

The Boulder Dam Association was a collection of boosters that originally formed in the Imperial Valley area in 1923. The group solicited membership and officers from the cities and civic bodies of the Southwest. On 1 May 1923, John Bacon, then Mayor of San Diego, wrote a letter to all area mayors, the trustees of banks, members and officials of Chambers of Commerce and county Farm Bureaus. The letter suggested that

> Conditions in connection with the deferred construction of the Boulder Canyon Dam on the Colorado River have reached such a stage that some definite planning is necessary before Congress meets. This is probably the most important project before the Southwest today. A meeting is hereby called . . . for the discussion of this problem and the adoption of definite plans and procedures as far as possible. A co-operative plan for participation will be presented by the Honorable Phil D. Swing, Member of Congress.

The members of this group were all dedicated to the building of Boulder Dam, and realized the "necessity for the distribution of facts and the need for prompt action" around the issue of the Boulder Canyon Project (Boulder Dam Association 1923: 11). The group gained important civic officials, including the mayors of San Diego and Riverside, supervisors of Orange, Riverside, San Bernardino, and Imperial counties, as well as Presidents of several county Farm Bureaus, Water Districts, and Irrigation Districts. The Association was thus made up largely of agricultural and political elites from southern California.

The group lobbied heavily for passage of the Swing-Johnson bill by printing and distributing pamphlets, writing press releases, making speeches to other civic associations, and contacting members of Congress.

Phil Swing was an important member of this group, and the Congressman worked closely with the Association. As one member of the association, Les Feader of the Western News Service, suggested to Swing in a letter on 3 May 1923 regarding the formation of the group: "There should be quantities of general information, drafted impartially and directed constructively toward one aim: that being THE BUILDING OF THE BOULDER CANYON DAM BY THE UNITESD STATES GOVERNMENT TO ITS FULLEST ECONOMIC HEIGHT." The Association was thus born, and its actions were well coordinated with Swing's legislative needs.

Such coordination can be seen in the editing process for a pamphlet "Boulder Dam and All American Canal as provided for in the Swing-Johnson bill: Flood Control, Irrigation, Reclamation, Hydro Electric Power, Homes for ex-Service men"[13] which Swing helped to edit:

> Humanity calls upon every red blooded American citizen to support this deserving measure if for no other reason [than flood control]—and relief can be granted without increasing taxation.

According to the Boulder Dam Association, then, the main goal of the Boulder Dam proposal was to control the flooding of the Colorado River. This could be done, the pamphlet argued, with financing through the sale of hydropower, thereby avoiding any increase in taxation. The Association was interested in boosting the dam through a cost-benefit analysis that explicitly compared the needs of "humanity" to a low cost program that would not increase the weight of taxes on the public. The nationalistic rhetoric of this appeal to humanity and morality ("every red blooded American citizen") implies that anyone who does not support the program is not worthy to call themselves American.

In addition to the flood control goal of the project, the pamphlet described the bonus found in the cultivation of new lands. The pamphlet argued that the increase of farmland was not the main goal, but an ancillary one:

> The proposed canal did not originate from a desire to bring additional lands under cultivation but from two other serious causes [flood control and avoiding a canal through Mexico] making it an imperative part of any program that may be adopted for the lower Colorado River. The additional lands are merely a decided advantage and supply one opportunity for Congress to compensate the ex-service men who offered their sacrifice in time of need.

In this appeal to ex-servicemen, the Association showed the direct benefits that the Boulder Dam project would grant them. New lands would be offered first to ex-servicemen in recognition of their "sacrifice." Opposing the project's aim to create new farmland opportunities for ex-servicemen was deemed unpatriotic and unappreciative of the sacrifice of the soldiers. Such selfishness on behalf of any opponents would be doubly suspect in the face of the soldier's sacrifice. The Association was thus interested in taking a position of moral superiority in relation to any opposition, while simultaneously showing the advantages of the project for the entire Southwest. In many pamphlets and circulars the Association argued that these benefits would be further extended to much of America (1923; 1926; 1927a; 1927b; 1927c; 1927d).

In September of 1927 the Association printed an important pamphlet for the boosting of the project, called "The Federal Government's Colorado River Project" (Boulder Dam Association 1927b). In the pamphlet the Association argued the merits of the project as they affected all of America; the subtitle emphasized the low costs (the project is "self-financing") for the "control and utilization of the flood waters of the Colorado River." The significance of this pamphlet lies in its timing and intricate arguments aimed largely at members of Congress.

In the introduction to the pamphlet the authors summarize the reasons for the project in typical and straightforward rhetoric. The project guards lands from flooding, and this "waste" floodwater will be "put to use" for domestic water, irrigation, and will be paid for by the production of hydropower. Overall, these reasons were put forward in the context of calling the project "a work of national importance" (Boulder Dam Association 1927d: 5).

The introduction then turns to the recent history of the project:

> The tense, dramatic ten days at the close of the last Congress, when the Swing-Johnson bill authorizing the project was kept continuously before the U.S. Senate, in the face of a determined filibuster in which power and politics were curiously blended, focused the eyes of the nation upon the great danger to the Imperial Valley and upon the titanic struggle going on unceasingly over the manner of the development of the Colorado River—one of the nation's great remaining natural resources (1927d: 5).

The Colorado River was discursively constructed here as one of the last "great natural resources" in America. Developing it involves a "titanic struggle" over powerful political forces. These interests, characterized as "great and powerful financial and landed interests" attempted to stymie the project through filibuster, favoring instead a plan that would "secure to their holding [in Mexico] an abundant and perpetual water supply for irrigation." These landed interests were led by Harry Chandler, owner of extensive tracts of land in Mexico and publisher of the *Los Angeles Times,* in league with private power concerns who wanted access to the power for their own gain (these oppositional discourses are discussed in the next section). The authors of this pamphlet argued that such selfish motives should not stand in the way of a development of tremendous national importance.

> Neither of these interests is satisfied with the plan of development recommended by Government engineers outlined in the Swing-Johnson bill: (1) Because this plan has regard primarily to American interests; (2) Because, in a most practical and effective way, the Swing-Johnson bill prevents monopolization of the hydro electric resources of the river by private interests. *The plan of the project and of the legislation authorizing it represents a determination that this great national resource must and shall be developed in the public interest and in the public interest alone* (1927d: 5, emphasis in original).

Not only was the plan of national importance, the bill was in line with Progressive-Era notions about the public interest. The development of this "great natural resource" was opposed because of the provisions against monopolization of benefits and provisions for American benefit at the exclusion of the benefit of foreign interests. The rhetoric of this pamphlet also focused on the practicality and efficacy of the project, reflecting the Progressive ideas of rational application of science to pragmatically solve problems of efficiency.

The benefits of the project were further propounded in its accomplishment of multiple goals.

> The project is wisely shaped to accomplish a variety of purposes. 1. It will forever end the flood menace of the lower river. . . . 2. It will put an end to an intolerable situation which now exists in Imperial Valley, due to the fact that the valley now secures its sole water supply by a canal running for some sixty miles through Mexico. . . . 3. Flood waters conserved at the dam will provide not only for irrigation needs below the dam in Arizona and California, but also for the domestic requirements of many rapidly growing cities of southern California. . . . 4. The dam and reservoir will incidentally create a large amount of hydro-electric power, from the disposal of which the project will be, in main, financed. Under the operation of the project, the flow of the river below the dam will be regulated and tendered comparatively uniform. Thus regulated, the river will be susceptible to use as a highway of commerce, whereas, at present, in its unregulated condition, it is of no practical value for this purpose (1927d: 9).

The Association's pamphlet argued forcefully for the regulation of the river. The "flood menace" would be ended forever. The project would solve the "intolerable situation" of a canal through Mexico with the All American Canal. Water for irrigation would be provided to multiple states, and drinking water would be made available to the "rapidly growing cities" that needed it. The national importance of this project was not only through the benefits gained by the areas of the West, but also through the improvement of transportation to the Western markets via this new "highway of commerce." These benefits should be gained through state action, yet the use of federal funds would be repaid with hydropower sales. The authors go on for more than 20 pages to argue their case. This pamphlet, then, made the case most extensively, and presents the set of arguments in the clear and forceful rhetoric of national interest. They followed the goal set out by Les Feader at the group's formation; their rhetoric was clear and took on the air of impartiality. It was, however, far from impartial. It was, as Feader suggested it should be, "directed toward one aim:" the construction of Boulder Dam.

The arguments of the Association resonated with many groups, especially ex-servicemen and the American Legion, but also women's clubs, Rotarians, and many others. Consequently, these groups also took on many of the arguments and the rhetorical style of the Boulder Dam Association.

Veteran's groups lobbied hard for Boulder Dam, arguing that benefits could flow to ex-soldiers. In order to gain these homes and farms for servicemen, the American Legion and other groups supported Boulder Dam and repeated the discourses on nature that constructed the river as an economic resource.

The national leadership of the American Legion supported the Boulder Canyon Project in an endorsement passed on 10 October 1923:

> WHEREAS the undeveloped resources of the Colorado River constitute one of our Nation's greatest assets, and
>
> WHEREAS the life-giving waters and potential energy of said river equivalent in value to millions of dollars is annually running to waste in the flood waters of the river and constituting a menace to life and property while it runs to waste, and . . .
>
> WHEREAS under said project thousands of ex-servicemen and women will be given preferential right to acquire homes and farms by entry upon several hundred thousand acres of public lands in Arizona, Nevada, and California, now arid and worthless, but which by this project will be made intensely productive and valuable,

> NOW THEREFORE BE IT RESOVLED that the American Legion in National Convention assembled does hereby favor the adoption by congress of a program for the ultimate complete development of the Colorado River.

The Legionnaires used many of the same arguments as the other boosters: the River was discursively constructed as a natural resource and a menace. The river was a "great national resource" whose "life-giving" waters, worth "millions" of dollars are going to "waste." The American Legion voted to support the "complete development" of the Colorado so as to secure the abundance of the "intensively productive and valuable" land that would be made available to ex-servicemen and women. The Legion's understanding of the River as an economic resource provides the justification for their call to fully develop it for veterans. Their political interest is clear, as their constituent group would be one of the main beneficiaries of the project. But the group also made the imperialistic argument that the river was a national asset, meant to be developed for the greater good and productivity of the nation, not just for the benefit of the ex-servicemen and women.

Secretary of Interior Albert B. Fall visited the region in December of 1921 as part of his research for the report that would be published in 1922, known as the Fall-Davis Report. Among many other official duties, he met with groups and held conferences on the Boulder Dam project. In his meeting with the American Legion, as reported in the Calexico *Chronicle* 9 December 1921, Secretary Fall said to the Legionnaires: "I know where you stand and I am certainly for you and with your help I believe we will put this over." Supporters of the project in the Imperial Valley were ecstatic over the visit from the Secretary, and gathered the various civic groups together for shows of support.

To publicize this united front, Mark Rose wrote, as reported in the Holtville *Tribune* on 20 April 1922, that the "American Legion, Farm Bureau, and all interests of Imperial Valley have unqualifiedly united in support of all provision of the Fall report. . . . [the] American Legion wishes public lands under All American Canal recommendation. Farmers of Imperial Valley wish flood protection by means of dam at Boulder Canyon sufficient to store a maximum yearly flow of river; also freedom from water shortage which can be secured only by new canal connecting Laguna Dam. All united in demand for high dam at Boulder because expense can only be borne by production of power."

Following shortly on Secretary Fall's visit, George P. Clements, manager of the Agricultural Department of the Los Angeles Chamber of Commerce, wrote a guest editorial in the *Los Angeles Examiner* on 22 June 1922:

> For twenty years the control and utilization of the Colorado River has been a common topic in Congress. For a much longer period the pioneer has taken the risk in development of Colorado delta lands, feeling that a righteous government would take recognition of their needs when shown through actual demonstration what potential wealth in revenue and production these lands under development offered the nation. Gambling against the hope of recognition of their government, these pioneers have risked their all in this endeavor. Into this project went their private capital, and they have bonded their lands, staked their credit to the last cent, and mortgaged the future generation upon the political promise of governmental aid in a systematic general control of this great unruly river which offered under harness greater national wealth than any other single natural resource in America.

Clements argued that the Boulder Canyon project is one of national importance. Brave pioneers have risked their lives, property, and future—in short, "their all"—for the foundations of this project. It was up to the federal government to be "righteous" and recognize these risks. The governmental aid to "control" the "great unruly river" that these pioneers gambled their future on is the nation's moral and ethical obligation. Clements' comments strongly exemplified imperialistic modernist rhetoric. The path of progress was clear to Clements, the rational application of science and engineering can solve problems of development and fulfill the promise that the brave pioneers wagered on.

Other endorsements of the Boulder Canyon Project include the Spanish War Veterans of Southern California and the Southern District Federation of Women's Clubs of California, both in January of 1924. By the end of 1924 a Boulder Dam Association flyer claimed that almost all of the Chambers of Commerce of southern California, the municipal Boards of Supervisors, Farm Bureaus and other "patriotic" and "sundry" organizations (by which the flyer apparently meant groups like the Imperial Irrigation District, the Lions, the Rotarians, the Women's Clubs, and etc.) had signed on to support the Boulder Dam Association.[14]

In other areas of the Southwest, especially closer to the site of Boulder Dam, boosters enthusiastically supported the bill. Anson Smith of Kingman, Arizona, publisher of the Kingman County *Miner,* and Charles Squires, of Las Vegas, Nevada, publisher of the Las Vegas *Age* were both tireless boosters of Boulder Dam. The two publishers reported on the proposal and construction of the dam on a weekly or bi-weekly basis. Squires was also active in a booster group called the League of the Southwest, an organization of civic leaders from multiple adjacent states. Squires traveled the Southwest boosting the dam and reporting its progress in his newspaper.

In a 1926 speech to the Friday Morning Club (a Los Angeles women's organization), Squires used state-centered imperialistic rhetoric to describe the ways that Boulder Dam will help develop the Southwest:

> Nevada looks to the Boulder Canyon Project as an essential aid to her future advancement. The consummation of that enterprise will bring to us, especially to Las Vegas and the adjacent country, an assured and prosperous future. It will irrigate our lands, bringing new life to our agricultural and horticultural districts. It will, by means of cheap hydro-electric power, revitalize our mining interests. It will attract wealth, population and industries upon which our future depends. It will establish Nevada as a great and prosperous state contributing of her natural resources to the prosperity of the entire southwest.[15]

Squires used many of the main strands of the discourse of imperial modernism: development of irrigation, hydropower, and adding the development of mining, and industry in general. But, Squires continued, the delays in Congress are harming the advancement toward the "prosperous future":

> By delay the tremendous potential values of the River will continue to waste to the sea; the growth of the entire southwest will be throttled; the threat of destructive floods will continue to lie like a pall over the lives and property of the Imperial Valley.

Squires' rhetoric leans towards the apocalyptic. He argued forcefully to many different groups in a similar vein, but his newspaper reached a much wider audience with his editorials and reporting.

In his newspaper the Las Vegas *Age* Squires continually boosted the Dam, using the same rhetoric as he did in his speeches. His articles carried headlines such as 17 January 1920: "HARNESSING OF THE COLORADO RIVER MAY BE ACHIEVED BY ACT OF CONGRESS: HOUSE BILL WOULD UTILIZE GIGANTIC FORCE THROUGH RESERVOIRS, FURNISHING ELECTRIC POWER, AND WATER FOR IRRIGATION: Las Vegas, Clark County and the Whole Southwest Will reap Wonderful Prosperity Should the Great Project Be Pushed to Completion as Now Probable." By using multiple lines of all capital letters, Squires put great force behind his boosting of the project. His rhetoric ("harnessing" the "gigantic force" to "reap wonderful prosperity") tends towards the bombastic.

In the text for one editorial, "The Unharnessed Giant of the Southwest," Squires described the Colorado River and the explorations by John Wesley Powell:

> But the spirit of Major Powell continued its work after his death through Arthur Powell Davis, nephew of the great explorer. . . . So, it is seen, the development of the tremendous power of the Colorado River, instead of being comparatively local, is . . . National in its importance.

Squires argued that this was no small, local project. Instead it was a great national project to build an empire in the southwest, which would bring prosperity and civilization to the area.

This argument, regarding benefits to the nation, was taken up by Las Vegas *Age* reporter Charles Benson on 3 July 1920.

> The benefits of this section [of the nation] to be derived from operation of the project are the permanent and solid basis on which our future growth and wealth would be founded. . . . As a matter fact, considering the power and storage to be developed, the Boulder Canyon project would pay for itself in increased wealth of the nation within four or five years after completion, and within ten years will be adding to the wealth of the Southwest a sum at least equal to its first cost; and this addition would increase annually.

Benson's article used the imperialistic modernist discourse, highlighting the industrial growth and production of wealth that would accrue to the city, the state, the region, and the nation. The project will pay for itself many times over from these benefits.

Later that summer, on 7 August 1920, the *Age* reported that several nationally-known engineers endorsed the project:

> A new chapter in the epic of the transformation of the 'Great American Desert'—a chapter surpassing in magnitude and daring every previous inroad of modern engineering in the desert country—is beginning to take definite form in the opinion of engineering and irrigation experts. . . . The discussion of the project was completely devoid of either oratory or critical debate and men who took part have lived or are living in the great drama of the West in its fight against the desert.

The domination of nature through the application of engineering technology and knowledge so as to produce this great wealth that would transform the Southwest is quite apparent in this article. This new "chapter" in the story of the transformation of the landscape is one of "great drama" that was told through engineering and irrigation experts. The project was deemed to surpass anything done previously in the region.

The Las Vegas *Age* also reported on the community reactions to the project on 22 October 1921 in a double size font: "LAS VEGAS-BOULDER

CANYON DAM RECEIVES ENTHUSIASTIC SUPPORT: THE CHAMBER OF COMMERCE HAS ROUSING MEETING AS A PRELUDE TO THEIR BIG DRIVE." The article described prominent locals who urged community action:

> Judge H. M. Lillis . . . spoke of the magnitude of the project and the need of the earnest support of every one in the community.

Community support for the dam was indeed wide and enthusiastic. By the time the project was finally approved in late December 1928, a group of over 400 Las Vegas residents made a pilgrimage of 30 miles to the dam site to give thanks (see Figure 4.6).

This vibrant civil society boosted the dam and canal project, keeping the idea alive in the face of the congressional and elite opposition described later in this chapter. The varied groups all used similar discourses about the domination of nature for the benefit of society. This benefit took the specific form of an empire in the southwest, which was portrayed as a gain for the entire nation. The project was also portrayed as morally and ethically sound, as it helped provide for ex-servicemen, protected

Figure 4.6 Washington D.C. *Star* 6 January 1929. Four hundred Las Vegas residents, kneeling in prayer near the construction site, give thanks for Boulder Dam.

against flooding, and the benefits of the project were all in the public and national interest.

This discourse of imperialism and national importance was used by many groups in boosting the dam, including Executive supporters, members of Congress, the engineers of the Bureau of Reclamation, as well as the national press. The discourse at the national level was largely similar to its usage at the local level, with a significant change in audience and legitimacy.

Executive supporters

The fight to win approval for Boulder Dam largely occurred in congressional committees, as the appropriate committee must approve all bills before a floor vote. Presidential views regarding such bills can help or hinder this process; however, in the case of Boulder Dam the continual, though sometimes unenthusiastic, support of a series of Presidents did not help coax the bill into law. The bill was first proposed under President Harding, who mildly supported the bill. Harding's Secretary of Commerce Herbert Hoover was integral in helping to negotiate the Colorado River Compact, and was a strong supporter of the dam and of river development in general. When Coolidge took office at the time of Harding's death, "Silent" Cal's response to the bill in Congress was moderate: while he was not totally opposed to the project he also did not push for its passage. The appropriation bill was finally passed at the end of Coolidge's administration, and construction was begun under President Hoover. Hoover's support was, however, somewhat hindered by the onset of the Great Depression. Franklin D. Roosevelt (FDR), on the other hand, made Boulder Dam and Grand Coulee Dam centerpieces of his New Deal programs to "prime the pump" of the economy. This fluctuating support from the executive branch helped prolong the debates in Congress for some time. By the time FDR came into office, Boulder Dam had been approved, but FDR's enthusiastic support helped push the dam to completion two years ahead of schedule and fourteen million dollars under budget.

Calvin Coolidge

Coolidge's unenthusiastic response to the Boulder Dam bill must be understood in the context of his general opposition to government aid to farmers. The result of this general opposition was a Republican-Democrat coalition made up largely of Westerners, a coalition that Roosevelt would build upon. Coolidge's veto of the McNary-Haugen farm relief bill in 1927 and another bill that would have increased the payments to ex-servicemen showed the president's general disposition towards government projects such as Boulder Dam.

Coolidge did, however, support the flood control feature of the dam, and commented, in his 1925 annual message (State of the Nation) that

> It is believed that the Federal Government should continue to be the agency for planning and constructing the great undertakings needed to regulate and bring into use the rivers of the West, many of which are interstate in character.

Coolidge's remarks showed a mild support for the project as a whole. Though he mentions regulation first, he does also support the "use" of the river. What he left out are the specifics of what such use would be. While other commentators wax eloquent about the multiple uses of the river, Coolidge's laconic comments showed his lack of zeal for the project.

Herbert Hoover

Herbert Hoover played an important role in the creation of Boulder Dam. Before his presidential term, as a young engineer and politician, Hoover was appointed Chairman of the Colorado River Commission while he was President Harding's Secretary of Commerce. Using his position, Hoover helped craft an interstate agreement (though not an uncontested one) on the allocation of water resources. This Compact provided the legal basis for the construction of the high dams that would store water for distribution to municipalities and agricultural fields. At the signing of the Colorado River Compact, Hoover (then Secretary of Commerce) made a speech, as recorded in the Arizona *Republican* (Phoenix) on 10 December 1922.

> So that the problems are not alone the problems of storage for the purpose of further expansion in power and irrigation, but there is also an urgent necessity for the erection of these great engineering works in order to protect and to remove the present peril to life and property of a large and fine colony of Americans in your own state [Arizona] and in the state of California.

Hoover used many of the same discursive strategies that the various boosters in California promulgated. Storage for irrigation and domestic uses were important, but control of flooding to protect the "colonies" prompted "urgent" and necessary action. The rhetoric of empire building can be seen in Hoover's choice of "colony" to designate the developments in the Imperial Valley and elsewhere in California and Arizona.

Hoover also wrote an article for *The Nation's Business* called "Conquering an American Nile" in August of 1922, as the Compact was being finalized. Citing the purposes of flood control, irrigation, and domestic use, Hoover argued for a federally-built dam at Boulder Canyon. Summing up Hoover noted:

> Perhaps the feature of the Colorado River proposal with the strongest appeal to the popular imagination is that it embodies an effort to apply a definite program, conceived solely in the public interest, of rational public control and allocation of benefits to one of the greatest unappropriated natural assets of the nation. Instead of leaving them to grab-bag distribution, as with most of our resources hitherto, it is proposed to plan the Colorado River for the public and to plan it without foolish plunder or foolish socialism.

Hoover's article embodied all of the main aspects of imperialistic discourse. He emphasized the nation-building benefits of the project, the characteristic Progressive-Era notions of rationality, efficiency, and development in the public interest. Interestingly he advocated planning on the scale of the entire river, an idea that Roosevelt would later expand upon—and for which he would be labeled as socialist.

Franklin D. Roosevelt and Harold Ickes

Although Franklin D. Roosevelt could not take credit for passing the legislation or beginning work on the dam, he and his Secretary of Interior Harold Ickes did take an active part in overseeing the construction of Boulder. As Ickes noted in an unpublished manuscript enumerating the benefits gained from Public Works Administration (PWA) programs, "We claim no credit for the conception either of Boulder Dam or TVA. But we hurried Boulder Dam to completion after we came in in 1933 and had it finished two years ahead of time. The power being generated there is now indispensable."

FDR's support of the dam was strong, but responsibility was largely delegated to Ickes for many such areas of domestic affairs. Ickes wrote dozens of articles in national magazines and newspapers to help garner support for the New Deal. He traveled all over the West, making speeches and stumping for support. Much of his time was taken up with projects that were started under the Roosevelt administration, but Boulder was of special importance to him due to its monumental character.

Ickes' relationship to the dam and his understanding of society's relation to nature was more complex than most of his contemporaries, as noted earlier. Many of his ideas do not easily fit within Scott's notion of a high modernist. While Ickes supported the building of Boulder Dam, Bonneville Dam, and eventually Grand Coulee Dam, he also recognized that even such magnificent structures were destructive in ways. The subtlety of his thought is captured in his notion of the "cooperation" between humans and nature, which, for Ickes, Boulder Dam characterized.

At the dedication of Boulder Dam on 30 September 1935, Ickes spoke of the glory of the dam and the benefits that would accrue to the nation because of it. His speech also candidly talks of the necessary domination of nature by humans:[16]

> Pridefully, man acclaims his conquest over nature. We have taught ourselves to believe that with our wit and with our might we have wrested from earth, sea and sky the necessities, the comforts and the luxuries of a complicated human civilization.

Ickes attacks the hubris of humans in the face of an unconquerable nature, and in this expresses a sentiment that is distinctly not high modernist. He propounded the ideas of conservation and efficiency in the use of resources that were the hallmarks of the Roosevelt administrations, putting Boulder Dam in pride of place among many achievements of rationality and efficacy. As noted in an earlier section, Ickes' labeled his notion of conservation as "cooperation" between nature and society. Ickes understood cooperation as a utilitarian idea:

> Through bitter experience we are learning the lesson that it is not through the destruction of nature or even though its complete subjugation that we will be able to serve our own greatest good; that if we would avoid the tragic mistakes of the past and assure for ourselves and for the generations that are to follow us the material and spiritual benefits which will flow from a wise and prudent use of the rich resources of nature, our approach must be in a spirit of cooperation.

Ickes' notion of cooperation between humans and nature is similar to Marx's notion of the cooperation between capital and labor. As Marx pointed out in *Capital* in the sections on cooperation, there is a dialectical dance between labor and capital (and now one would add the state) in the process of working out how to optimize the political-economic system (1867/1967: 333). Ickes argued that humans and nature similarly cooperate, solely for the benefit of human society. Ickes believed that Boulder Dam was a prime example of such cooperation. The idea of cooperation, which usually implies a purposive mutuality, appears closer to complete subjugation than Ickes let on in his definition, and here his high modernist tendencies become more apparent. In the speech, Ickes argued that "no better example" of this cooperation could be found than Boulder Dam, a structure that, to many, symbolizes the human domination of nature more fully than perhaps any other example (McPhee 1971: 158). Ickes believed that human domination of nature was justified only to the extent that the benefits could be distributed according to his guiding notion of the "greatest good for the greatest number." This domination, according to Ickes, was most properly done by the hand of government in order to benefit all society. His dedication speech continued:

> I venture to hope that this dam, with its great storage of health and wealth and happiness for thousands of people, will stand as the definite opening of a new era with respect to the natural resources of America—an era of conservation which means the prudent use of all our natural resources for the greatest good of the greatest number of our people; an era that will recognize the principle that the riches of forest and mine and water were not bestowed by God to be ruthlessly exploited in order to enhance the wealth of a small group of rugged individualists, but were beneficently given to us as an endowment to be carefully used for the benefit of all the people. On no other theory would the Federal Government be justified in so generously opening the doors of its treasure house for the building of this and other similar projects that will turn large sections of this breathtaking Western country into rich homesteads where a happy and contented people will find it possible to live those comfortable and worth-while lives that we covet for every man, woman and child in these United States.

FDR also subscribed to this utilitarian notion. His speech at the Boulder Dam dedication in September of 1935 celebrated the victory of humans over a desolate environment. His speech is marked by the recurrent ideas of the imperialistic discourse: state building and national benefits to the project—especially irrigation and hydropower—as well as the control of the river from flooding.

A new purpose was added in FDR's speech, an idea that had not been discussed in the debates over the construction of the dam and an element that had not entered into the engineer's and economist's cost-benefit analyses: recreation.

> Ten years ago the place where we are gathered was an unpeopled, forbidding desert. In the bottom of a gloomy canyon . . . flowed a turbulent, dangerous river. . . . We are here to celebrate the completion of the highest dam in the world, a dam which rises 726 feet above bed rock of the river and alters the geography of a whole region. . . . Beautiful and

> great as this structure is, it now ought to be considered less as an engineering masterpiece and more in its relationship to the industrial development and in its contribution to the health and comfort of the people who live in the Southwest. . . . For a generation the people of the Imperial Valley had lived in the shadow of disaster from the river which provided their livelihood, and which is the foundation of their hopes for themselves and their children. . . . A gigantic playground has been created in barren and mountainous territory sadly in need of recreational facilities. . . . The wealth created by the construction of Boulder Dam is a national asset.

FDR touched on each of the main themes of the imperialistic discourse regarding nature and society in the American West. The region in which the dam was built was discursively constructed as a wasteland: the "forbidding" place of a "turbulent" and "dangerous" river where people dared not go. Boulder Dam changed the geography of the whole region, not only its physical geography, but also its economic and human geography. Boulder Dam brought a playground to the Southwest as well as wealth, health, and comfort. All of this amounts to Boulder Dam as a "national asset."

Franklin D. Roosevelt and Harold Ickes celebrated the completion of Boulder Dam as one of the crowning moments of the New Deal promise to reinvigorate American industry. As newspapers such as the San Bernardino *Sun* had noted, Boulder represented an achievement greater than the building of the Panama Canal by his cousin Theodore Roosevelt; Americans had taken a personal interest in it, and the dedication of the dam was, in Roosevelt's words, "a climax in man's eternal contest with rugged nature."

Swing, Johnson, and Congress

As with all appropriations bills, the Boulder Dam bill had to be approved by Senate and House Committees before being passed by both houses and moving on to be signed by the President. The Boulder Dam bill was assigned to the Committee on Irrigation and Reclamation, and several hearings on the different versions were conducted and several floor speeches were recorded in the Congressional Record. Phil Swing's activities in relation to the Boulder Dam Association and other civil society groups dove-tailed with his work in Congress, where—as author of the bill—he cooperated with the Committee members and his co-author in the Senate, the Progressive Republican from California, Hiram Johnson. The same imperialistic discourse about national progress, the domination of nature, and the building of an empire in the

West that was seen in civil society could also be seen in the debates before the Committee and on the floor.

Such discourse was especially evident in the hearings before the Committee on Irrigation and Reclamation, which opened discussion with a reading of the bill and commentary by Phil Swing (House of Representatives Committee on Irrigation and Reclamation 1924c: 5). Swing noted the support from various civil society groups such as the American Farm Bureau, the American Federation of Labor, the General Federation of Women's Clubs and the United Spanish War Veterans. He argued that the American Legion was behind it—their periodical called it the "Biggest job in the world"—claiming that the benefits from the project "will, for the amount of money invested . . . return greater dividends and assets to the American people than any other project (House of Representatives Committee on Irrigation and Reclamation 1924a: 6). In short, Swing argued, the people of the nation were behind it, it made economic sense, and the committee should approve it. The hearings then read into the record a number of letters, including one from Mulford Windsor, President of the Arizona State Senate (House of Representatives Committee on Irrigation and Reclamation 1924a: 1):

> Let us drive directly to the heart of the matter. . . . I want to see the Colorado River utilized—waste ended. I want to see the third greatest waterway of the country comprehensively, systematically developed, its destructive forces subdued, their vast possibilities converted to the uses of man.

Windsor, who was in the midst of the imbroglio regarding Arizona's refusal to ratify the Colorado River Compact, expressed his desire to move on with development and supported the Compact and the Boulder Canyon Project bill. In language that leaves no doubt regarding his devotion to the principles of development, Windsor argued forcefully for the bill and the "comprehensive" and "systematic" development of the Colorado. The river was nothing if not a resource for human use.

Such language of domination is characteristic not only of the dam's proponents; its detractors, including Carl Hayden, U.S. Representative (larer senator) from Arizona and opponent of the dam and the Colorado River Compact, made use of almost identical rhetoric. Hayden felt that the water allocation was unjust (as will be explored in detail in the next section), but was fully in favor of development. Hayden commented to the committee that (House of Representatives Committee on Irrigation and Reclamation 1924a: 13):

> [W]e all recognize in the Colorado River the greatest single undeveloped resource in the Southwest if not in the Nation; that it is a menace to the lowlands along the lower river, . . . and that the prosperity of our part of the country would be greatly enhanced by the earliest possible development.

Thus even the opponents of the bill used the rhetoric of imperialism to discuss the benefits that developing the river would bring to the region and the nation.

The hearings continued in March 1924 with the reading of a letter by Secretary of Interior Hubert Work in which he described the project's importance to the nation in superlative terms (House of Representatives Committee on Irrigation and Reclamation 1924b):

> The time has arrived when the Government should decide whether it will proceed to convert this natural menace into a national resource. . . . The Colorado River is now a local menace. It may be converted into a public utility of first importance, particularly to the southwestern United States. Its possibilities rank with those of the Panama Canal, the St. Lawrence waterways, or the Territory of Alaska.

Work compared Boulder Dam favorably to successful projects that captured the nation's imagination over the past generation. Boulder Dam, Work argued, would rank with the greatest of national achievements. Though at other times Work's comments had been interpreted as urging caution, the overall tone of his rhetoric at this point was effusive. For example, in his letter to the committee he described the plans for the dam in comparison to other giant structures, noting that it dwarfed them all.

The hearings received other testimony, primarily oral, from many different parities, most of them enthusiastic about the dam proposal. In response to a question from Representative Little regarding the market for Boulder Dam power, H. A. Barre, executive engineer for Southern California Edison Company, stated:

> I think Los Angeles is going to grow reasonably. We have had a crazy growth in the last two years. These things cannot possibly continue. . . . For example, the last couple of years there has been a tremendous influx of the kind of population that is not self-supporting. People would come in driving overland in what was left of an automobile. They would have a few hundred dollars, and they would put it down on one of these small houses that have been scattered around the country so freely. Then they would get a job to pay for the house by building another house for the next fellow that was going to come along. Now, you cannot keep that up indefinitely. Los Angeles will now grow in direct proportion to the development of its industrial production.

Barre urged the construction of Boulder Dam so as to spur the development of Los Angeles and the productive power of the Southwest. The power would invigorate industrial growth and provide real jobs that could transform the "non-self-supporting" migrants. The building of such an empire, he argued, would provide for real growth instead of unsustainable growth that was largely speculative.

Arthur Powell Davis also made a presentation to the Committee. Davis answered many detailed engineering questions, and ended by echoing Secretary Work, saying that the Boulder Dam "will turn a natural menace into a natural asset."

Swing's colleagues, such as Senator Hiram Johnson, were staunch supporters of the bill. Johnson, the sponsor of the Senate version of the Boulder Canyon bill, made several speeches as well as entering pro-dam newspaper articles into the Congressional Record.

In a 7 July 1922 guest editorial for the *Los Angeles Herald,* "Boulder Canyon Dam; Project Means New Empire for So. California," Senator Johnson made many points that came straight from his speech texts:

> The development of the Colorado River Basin proposed in the legislation by Representative Phil D. Swing in which I have been very happy to cooperate, constitutes the greatest construction project before the American people today. Within it there is greater promise of material prosperity for California than in any pending legislation or in any even of recent years.

Johnson saw Boulder Dam as the foundation upon which California could build a "new empire" that promised great "material prosperity." He identified it as the greatest project pending or even thought of in many years, and as such it deserved the public's utmost attention and support. Its importance as an engine for industrial progress, he argued, concerned everyone in California. But the dam was not just the whimsy of California. Other states and the nation at large will benefit, Johnson argued:

> [The river's] course runs through a mountainous country where nature has created narrow, high-walled gorges, especially adapted as sites for great dams. It has demonstrated now that its waters can be controlled, and can be utilized and developed, and be made of inestimable service to this and all succeeding generations. The pending legislation is the mature judgment of the experts and engineers, both federal and state, for this control, development and use.

In this exemplary editorial, Johnson painted a portrait of the dam as being approved of from all sides: both federal and state experts and engineers, the public, and even nature. The great gorges of the West, Johnson argued, are adapted to dam building. Though it may simply be a rhetorical flourish, Johnson appears to argue that nature had provided dam building with an evolutionary advantage in the West. In many ways, then, the dam builders were simply completing nature's intentions and bringing prosperity and development to the Southwest in the process. Though not common in the congressional debates, this idea of nature supporting the endeavors of man (similar to Ickes' idea of cooperation) showed the discursive basis upon which the nature-society relationship was constructed. Nature approved of such projects, and in fact has even invited them. The gendered character of this rhetoric parallels many constructions of nature as feminine, nurturing, and supportive of human (male) undertakings, though also potentially fickle, as Pitkin argues *fortuna* can be.

As transcribed into the Congressional Record, Johnson's speeches reflected all of the rhetoric used in the discourse of imperialist modernism. In the debate on what would became the penultimate version of the bill, in April of 1928, Johnson noted the importance of the Boulder Canyon Project (Senate 1928: 7245):

> It is an important piece of legislation, first, because it is the greatest constructive work now pending in the country, the greatest constructive work of our generation; and secondly, and equally important, because it is the test in the Congress of the United States of whether there are the power and the courage in the Congress, in the face of great private interests and enormous amounts of wealth, to enact legislation which is required by a part of the people of the Nation.

The newspapers commented on how Johnson characterized the bill as a challenge to Congress (Squires 1928). His Progressive insinuation was that Congress must be courageous and stand up to the corruption of government serving the business elites who wanted to stop this project.

Johnson's speeches also showed the imperialistic discourse of domination, in this case tinged with wonder (Senate 1928: 9437):

> [W]e are going to erect by the Government a dam that ought to challenge the admiration and fire the imagination of every man who has anything to do with it.

In his comments on the Senate Chamber floor, Johnson constructed the debate as one in which any member of the opposition was guilty of a lack of imagination. His rhetoric contained many of the strands of the imperialistic discourse, such as this moral superiority. Elsewhere Johnson constructed the river as violent and in need of control by humans, describing the river's "torrential moods" in which "it has torn in Mother Earth great chasms" (Senate 1928: 7245). Overall, Johnson's comments in the Senate exemplified a modernist perspective that was heavily Progressive in its understanding of the role of government, a role that imagines an imperialistic extension of social and political control over territory and population.

In February of 1927 *The Congressional Digest* devoted an entire issue to the Boulder Dam bill. The issue included pro and con arguments on all aspects of the bill, from construction in general to the power question. The most important actors were included in the printed debates, including Phil Swing, Hiram Johnson, and Arizona Representative Carl Hayden.

Secretary Work's comments opened the issue: "The Importance of Developing the Lower Colorado" (Work 1927). Work argued that the "recurring threat of disaster" from floods required action. Proudly noting that in its brief history, "the United States stands foremost in the damming of rivers and harnessing their forces for the benefit of mankind," Work suggested that the uses of the river will be irrigation and drinking water as well as flood control. Most importantly, however, were the "industrial uses" that will come from the "electric energy, which when disposed of, will pay the entire cost of development" (Work 1927: 39). Work outlined the arguments for construction, but also noted the special role of nature. "Nature," Work argued, "placed this canyon about where it ought to be of the proposed purposes" (Work 1927: 39). Similar in sentiment to Johnson, noted above, Work clearly thought that nature intended a dam for in this location.

The issue of *The Congressional Digest* continued with comments from Commissioner of Reclamation Elwood Mead and an analysis and explication of the bill text and the history of the area. The debates over the central issues were taken up by both Senators and Congressmen, but all agreed that the development must occur (see the next section for detailed discussion of such debates). Even opponents of the bill argued that the river must be fully developed; they simply disagreed over the form of the dam and the allocation of its benefits.

The discussion within the House and Senate were characterized by a general agreement regarding the appropriateness of dominating nature. Not only would the dam produce many benefits for society—both locally and nationally—but it also appeared to be intended by nature and the evolutionary process. Boulder Dam would bring a new era of prosperity and industrial development to the Southwest; a new empire that would ensure the predominance of the United States in the world. Boulder Dam was favorably compared to the other great projects and acquisitions over the previous 30 years—the

Panama Canal, the St. Lawrence Seaway, the Alaska Territory—and was thus given the legitimacy that comes with such associations. In short, Boulder Dam was seen as essential to the building of a new empire—and therefore to America's continued industrial and economic progress and prosperity.

The Bureau

The Bureau of Reclamation played a strong role in boosting Boulder Dam, but in the early 1920s the Bureau's reputation was suffering due to a lack of repayments for irrigation and reclamation projects. The original 1902 law required that projects be paid for in part by recipients of water, and most farmers had defaulted. Commissioner of Reclamation A. P. Davis had been developing a plan for the development of the Colorado River for some time, sending his engineers and surveyors to sites throughout the Basin. By 1923 when the Reclamation Service was reorganized and renamed the Bureau of Reclamation, Davis searched for a large-scale project that could help him salvage his reputation. Phil Swing's election to Congress and drafting of the Boulder Canyon Act with Hiram Johnson was just such an opportunity for Davis, but it came too late to save his career. Davis was replaced by Elwood Mead as Commissioner of Reclamation in 1924, and Mead would be the man who won the reputation for building the dam (the reservoir backed up by Boulder is named Lake Mead)—though it was primarily Davis who planned Boulder and continued boosting the dam even after leaving public office.

Davis was co-author with Secretary of Interior Albert Fall of the 1922 report entitled "Development of the Imperial Valley," also known as the Fall-Davis Report. This was the first official recommendation of the construction of a dam "at or near Boulder Canyon" with a canal carrying water to the Imperial Valley that was entirely within the United States (Bureau of Reclamation 1922: 21). Davis was a tireless promoter of the dam, for he saw it as a way to save the Bureau with a stupendous engineering achievement that could be used to fight the Bureau's opponents in Congress. In speeches, articles, and correspondence, Davis boosted the Dam and the progress he thought it would bring to the Southwest and the nation. Davis was unsuccessful, however, at convincing his opponents in the administration of his plan to save the Bureau.

With the dismissal of Davis, Elwood Mead's first task was to "fix" the Bureau. He set out on a journey through the West to investigate the problems and promise of reclamation projects. Secretary of Interior Hubert Work accompanied Mead, and the Bureau produced a jocular account of the trip entitled "Studying Reclamation with Secretary Work and Doctor Mead" (Work and Mead 1926). The humorous subtitle indicates the light-hearted perspective that the piece takes, "Being a Chronicle of a Journey into the Dry and Sizzling West in Mid-Summer, 1925 To Diagnose the Ills of Federal Reclamation and Find and Apply an Effective Remedy." The book used a subtle humor in describing Federal Reclamation as "ill," "ailing," and the "various specialists called in had tried their pet theories, but the patient grew steadily worse" (Work and Mead 1926: 9). The figure accompanying this introduction, reproduced below as Figure 4.7, shows Work and Mead examining a decrepit patient on a gurney, wielding a machete rather than a scalpel.

Mead became a tireless booster of Reclamation in general and Boulder Dam in particular. Building on the rhetoric that had worked so well in proposing the dam and other projects, Mead elaborated the imperialistic and modernist discourse on reclamation in the West.

Mead's February 1929 press release is a rich document describing the history and purposes of Boulder Dam (Mead 1929b):

FOR IMMEDIATE RELEASE IN MORNING PAPERS OF FEBRUARY 21, 1929.

> Into this region which is the hottest and driest part of this country, the Colorado River brings the melting snows of the high mountains at its head. It is being used in a climate which makes irrigation necessary to healthful and comfortable living conditions and which gives marvelous variety and value to irrigated crops. These values have grown with

Figure 4.7 Operating on the Bureau of Reclamation, 1926.

> increase of people and advance of civilization. The water of the Colorado is needed on farms, in factories and in the homes of two million people now living in the Southwest and for the larger number of millions destined to live there soon.

Mead reflected many of the elements of the imperialistic discourse in his press release; indeed most elements of his rhetoric and his administration of the Bureau were essentially similar to that of his predecessor A. P. Davis, and the overlap in rhetoric with boosters like Swing is apparent. Such similarity in rhetoric was in part due to explicit coordination. In fact, Mead had shared this release with Phil Swing, who offered little comment on it.[18] In the press release, the project was praised for its multiple purposes of irrigation, hydropower production and flood control for the Imperial Valley, so named, Mead notes, for the "opulence of its crops."

The Imperial Valley, Mead explained, before irrigation was a "hideous, unpeopled waste, without streams or springs, with rainfall of only two to three inches a year, and so hot in summer that those who had to cross it did so at night" (Mead 1929b: 2–3). Boulder Dam solved this problem by bringing flood control, water, and power to the people, cities, and industry:

> The pathway of civilization in the arid lands follows the banks of streams. . . . Those who have followed most closely the history of this undertaking feel assured that the time is not far distant when this great national resource will be harnessed, when devastating floods will no longer menace the lives and fortunes of the people of the Imperial Valley, when an adequate domestic water supply will be available for the growing cities of the southwest, and when the hum of electric motors will tell the story of . . . industrial development (Mead 1929b: 17–18).

Mead's press release was alive with action: the river will be "harnessed," stopping the "devastating floods" thereby helping develop the "growing cities" and producing the "hum of electric motors." Such rhetoric of human domination was put in the context of national benefit as well as regional fortune, and is used over and over in Mead's work.

One article in particular, published in the *Review of Reviews* in September 1929 was entitled "Conquering the Colorado." In this article, rich with descriptions of the rugged, hot, dry, and imposing geography, Mead outlines the argument for a state building endeavor in the desert (Mead 1929a: 55):

> The pioneer idea of development was for each man to look out for himself and let the devil take the hindmost. Individuals and companies

> laimed rivers, and dug canals to divert them, without thinking what others were doing above and below on the same stream. We have reached the end of that kind of unrelated action. . . . [Boulder Dam] has enlisted the cooperation of the nation, states, cities, irrigation districts, and important hydroelectric corporations.

It was the responsibility of the federal government to enlist the cooperation of the many interests in order to get the job done. Mead argues that the Bureau of Reclamation, Swing, Johnson, and others have done just that, and united they have "conquered the Colorado."

In other venues Mead boosted Boulder Dam by showing the connection between industrial and economic development and the Boulder Canyon Project. In a speech to the Commonwealth Club of San Francisco on 10 July 1931, Mead described the influence on the Southwest of Boulder Dam, "the destiny of the Southwest will be shaped by building Hoover Dam.[19] If it is built, the cities of that region will continue to grow. If it is not built, they will stagnate." Mead's rhetoric equated the dam with continued economic and industrial progress, as well as population growth. As part of the public edifice of the Bureau of Reclamation, Mead presented the best of all possible perspectives on Boulder Dam.

It was also under the guidance of Mead that the Bureau of Reclamation produced a rich document used for educating members of Congress, new Executives, and the general public about the dam, the dam site, and the heroism of the Bureau's Engineers. Entitled "Illustrated Lecture on Boulder Canyon Project" (Bureau of Reclamation 1932), prepared by the Office of the Chief Engineer, the document described the history of the project, the geography of the region, and the design of the dam.

The Lecture opened with a general discussion of the region and the history of the project, outlining the reasons for the dam, highlighting flood control. The Colorado was called "silt-laden, sluggish, and unattractive" where it emerges into the Gulf of California (formerly called the Sea of Cortez), but this image was contrasted with the headwaters of the river in the high Rocky Mountains where it was called a "flashy stream" (Bureau of Reclamation 1932: 13). The "turbulent, dangerous force" of the Colorado necessitates Boulder Dam to protect the farms of the Imperial Valley, described as formerly a "forbidding waste whitened by skeletons of perished travelers" that has been transformed through irrigation into "one of the most productive regions of the United States" (Bureau of Reclamation 1932: 16), as exemplified in the slide which accompanied this portion of the lecture, reproduced in Figure 4.8 below.

Figure 4.8 Lower Colorado Farm Scene (slide 8), Bureau of Reclamation 1932.

The lecture proceeded to describe the flood danger that the Colorado poses to these magnificent lands, saying that "control of the River was lost" (Bureau of Reclamation 1932: 18), as shown in Figure 4.9. By using the passive voice, the authors de-emphasize the mistakes that Rockwood and the California Development Company made under pressure to keep water flowing to the Imperial Valley. With this justification, the lecture then discussed the canal's route through Mexico and the hardships that Imperial Valley farmers faced because of this. The Boulder Canyon Project, the authors maintained, would solve the problems of flow regulation and siltation that caused the original breach as well as providing for an All American Canal to remedy the Mexican "problem."

The solution to these many problems was found in the Boulder Canyon Project Act, written by Swing and Johnson, and passed under Coolidge's term of office, all of which figure in a prominent image presented during the lecture. The lecture amazingly passes over the extraordinary eight-year debate over this Act. Eager to describe the design features of the dam, the lecture skips the turbulent history of the proposal for the dam and instead moves directly to the award of construction to Six Companies and the location and engineering challenges posed by the geography of the region, as exemplified by the aerial views of Black Canyon, in Figures 4.10 and 4.11 (which includes a rendering of the dam) below. Such an elision of history, a skipping over of the controversy that the dam provoked, shows an imperialistic logic.

Figure 4.9 Colorado River Break, 1905 (slide 10), Bureau of Reclamation 1932.

Figures 4.10 and 4.11 Aerial views of Black Canyon, with an artist's conception of the completed dam (slides 22 and 23), Bureau of Reclamation 1932.

By leaving out contestation, the Bureau constructed a new history of the dam. In this history there were no Senate filibusters, no elite detractors, and no interstate rivalries. Most imperialist endeavors seek to erase opposition, and the Bureau's "Illustrated Lecture" is no different.

Using rhetoric that described the challenge and conquest of the river and the canyon—the heat, the cliffs, the falling rock—in terms very similar to those used by Clyde Ramer in his poem quoted as an epigram to this chapter, the Bureau lingers in great detail on all aspects of the dam's engineering design, artistic design, and the construction process.

Overall the Bureau's "Illustrated Lecture" served to present the best possible public face of both the Bureau and Boulder Dam. The lecture skipped over the political and economic geography of Boulder Dam, particularly the intense partisan squabbling that characterized the years of debate in Congress and the fierce opposition exhibited by local and regional elites (as is discussed in the next section). Instead, the Bureau deftly shifted the focus to the brilliant engineering solutions for challenges posed by the geography of the Colorado.

The Bureau also produced many pamphlets, brochures, and a project history that reiterated the benefits, challenges, and victories involved with constructing Boulder Dam. The public face presented by the Bureau was one that used the rhetoric of the brave worker described as a hero with a mission to conquer nature, the ingenious engineer who enabled the construction of something never done before, and the visionary boosters who pushed the project when detractors had lost faith in America's ability to conquer the Canyon and the River. This selective history of the project leaves out the acrimonious debates and partisan squabbling that marked the long road to building Boulder Dam.

A 1933 brochure described the dam as "more massive than the Pyramid of Cheops in Egypt" (Bureau of Reclamation 1933: 3); elsewhere Bureau pamphlets described it as "The World's Most Impressive Engineering Spectacle in the Center of an Area of Unsurpassed Scenic and Historical Interest" (Bureau of Reclamation 1936: 1). The Bureau presented the dam to the American (and international) audience as an achievement to be celebrated, as a conquest of humans over nature, and as an unmitigated positive accomplishment.

The boosters of Boulder Dam were thus primarily governmental engineers and administrators in the Bureau of Reclamation, members of Congress, and the farmers of the Imperial Valley. They overwhelmingly used a discourse of imperialist high modernism to explain and justify their project: Boulder Dam would provide a basis for expanding and completing an empire in the West. Boulder would be a monument to science, technology, progress, and human potential. To its detractors, however, Boulder Dam represented at best a waste of taxpayer money, and at worst an instance of "creeping socialism."

OPPOSITION TO BOULDER DAM: THE DEFENSIVE-MODERNIST DISCOURSE OF ELITE DETRACTORS

Opposition to Boulder Dam was primarily a reaction to the proposals of Swing and the Bureau. The opponents were modernist in their beliefs in rationality and progress, but they greatly distrusted the state and advocated instead for private control. In their view, the state could legitimately build a dam for flood control and perhaps irrigation, but to fund this through the public distribution of hydropower was unacceptable. The opponents argued vehemently against a state-run dam on the Colorado River. Their opposition, however, did not build from the discourses of resistance to Hetch Hetchy Dam; instead opposition arose from a self-interested and defensive position against state action.

By the mid 1920s when Swing and Johnson were pushing their bill in Congress, the Hetch Hetchy controversy had come to a close and John Muir and the Sierra Club had lost the battle. Perhaps this closing-off of effective contestation established a vacuum of opposition. Certainly there were no social movement entrepreneurs interested in the Colorado River as there had been in protecting Yosemite National Park and Hetch Hetchy Valley. Most clearly, opponents were created by a political system that threatened their interests. They thus simply acted to protect those interests in a defensive manner rather than act positively to protect nature. Such a level of causation is not easily determined, but the reality of opposition to Boulder Dam was that it was self-interested on the part of the political and economic elite of southern California, and defensive on the part of rival states in the Southwest.

The primary opposition to Boulder Dam came from elites such as Harry Chandler, owner of the *Los Angeles Times,* who controlled vast crop acreage in Mexico south of the Imperial Valley. His lands were watered from the Imperial Irrigation District's canal as it swung through Mexico. The Mexican government required access to this water in return for allowing the canal route through Mexico, and Chandler bought the water from the Mexican government at a very advantageous rate. An All American Canal would cut him off from this water supply and his lands would be unusable for agriculture. Chandler was a modernist in that he found state-run projects acceptable, but only to the extent that they did not negatively affect his own holdings. His opposition to the dam at Boulder was thus inextricable from his own self-interest.

Similarly, the state of Arizona opposed Boulder Dam due to its own understanding of its best interests. Arizona felt slighted by the disproportionate allocation of water in the 1922 Colorado River Compact and refused to ratify it for many years. Furthermore, State officials felt that a site at Glen Canyon, farther up the Colorado between Utah and Arizona, should be developed before Boulder. This dam site would make it more likely that water and hydropower would be delivered to the closer Arizona markets instead of Los Angeles. Arizona was thus not anti-modernist or even anti-statist. The state politicians were simply defending the state's perceived best interests.

These discourses of defensive modernism often shifted their focus away from directly self-interested arguments and towards the politically expedient arguments about equity and ethics; they are defensive stances that cloak themselves in the discourse of modernism. The self-interest was often covered over by a lofty discussion of science, rationality, and progress. While Arizona discussed equitable distribution of water resources and state's rights, Chandler and the National Utility Association claimed to present unbiased facts, even reprinting their opponents' statements and adding their own interpretations.

Discourses of modernism, in a very general sense, stress rationality, science and technology, historical progress, and a break from tradition and custom (Weber 1922/1946: 155). The discourse of opposition to Boulder Dam firmly stressed such discursive strategies while simultaneously downplaying any self-interest, though clearly motivated by such material gain. I thus term these discourses self-interested defensive modernism.

The Los Angeles Times: *Harry Chandler Opposes the Dam*

Harry Chandler, publisher of the *Los Angeles Times,* was the most vocal opponent of the Boulder Dam Project in the West. Though there were many elite detractors of the dam, Chandler led the fight, for his extensive resources allowed him to fight the project both through the distribution of pamphlets and in the pages of his newspaper. The *L.A. Times,* founded in 1881, was known in the first half of the twentieth century as an abysmal, yet powerful periodical. A new history of the *Times* was reviewed by Hendrik Hertzberg for *The New Yorker* (2001: 187–8):

> The *Los Angeles Times* was known for one thing above all: its badness. This badness was of no ordinary kind. The *L.A. Times* was venal, vicious, stupid, and dull. It was abominably written and poorly edited. It

> existed to advance the unfathomably reactionary political views of the family that controlled it and, even more, to aggrandize that family's private economic interests. . . . Its political coverage was propaganda. Its local coverage was a combination of gaseous boosterism and outright suppression of anything that might inconvenience the business community, especially those considerable sectors of it in which the paper's owners had a direct interest.

Chandler was a partner with several other elites who had reason to oppose Boulder Dam, most significantly the power companies and agricultural businesses. Chandler was thus central, and he fought the dam by opposing almost all of the reasons that the boosters supported it. Chandler argued about the extent of the flood menace to the Imperial Valley; he took exception to the public intrusion into the utility industry; and he pointed to the injustice of proceeding with the Colorado River Compact against the wishes of Arizona. Of particular importance to Chandler, as the owner of extensive agricultural lands in Mexico, was the lack of a treaty ensuring an adequate amount of Colorado River water flowing across the international border below the Imperial Valley.

One of Chandler's pamphlets (The Colorado River Land Company 1926),[20] entitled "The Federal Government's Colorado River Project," outlined these issues. Chandler recounted the many ideas for controlling the flow of the Colorado, and how each one had been found impracticable. Chandler then called the project "at most a paper scheme" (*ibid.:* 14), claiming that no core samples had been made to verify the legitimate possibility of building a dam at Black Canyon. The pamphlet then moves on to describe the inequities on the part of the state of Arizona, which did not sign the Colorado River Compact, and the disproportionate power on the part of California and the Upper Basin States.

"The Real Issue," according to the pamphlet was the "Control of Power" (*ibid.:* 17–18):

> [T]he real fight going on behind the scenes turned upon the question of whether or not the private power interests of the country were to eliminate from the legislation the provision inserted by the administration, for good reasons, for the protection of the financial integrity of the project, to secure fairness in allocation of benefits, and to prevent unnecessary construction costs.

Chandler purported to illuminate the real issue behind the rhetoric of the debate. He exposed the power interests battling with the federal government, who wished to finance the project through the sale of power. By exposing this "real" issue, Chandler decentered the debate from his own interests as a member of the economic elite who would benefit from the sale and distribution of the power by private companies.

Chandler went on to underscore the "urgent need for action" (*ibid.*: 17–18) based on the flood menace to the Imperial Valley. He suggested, however, that there was "some confusion as to the nature of the flood menace" (*ibid.*: 18). Every year that goes by, Chandler argued, "brings disaster that much closer." Urgent action was surely needed, but the various sides were locked in debate regarding power interests and water allocation disputes. These delays were "intolerable." Chandler argued that among the multiple purposes proposed for the river, the superior goal was flood control. He argued that it was only the federal government who could achieve this end: "only the strong hand of the government in full control of the works will assure the weaker but more important uses their proper treatment" (*ibid.*: 20). In order to achieve the crucial ends of flood control, the government should build the dam, and the interminable delays need to be overcome. This moment of statism is an anomaly within Chandler's rhetoric, and must be seen as support for statism only when he stood to benefit.

Chandler's rhetoric thus turned towards the modernist, calling on the government to act with its "strong hand" to bring progress as soon as possible, but only on the issue of flood control. Chandler's arguments brought forward the government's role as mediator, while making jabs at the booster's arguments about public power and water allocation. An international treaty was required, he argued, as was an equitable distribution of water. But these arguments were consuming precious time. Instead of solving insurmountable problems, he contended, the government should build flood control works and leave all else to private interests.

Chandler's pamphlet thus puts the emphasis on building the dam as soon as possible, allowing the debates to be resolved at a later time. This modernist rhetoric (put aside differences, be responsible, focus on progress and efficiency) allowed Chandler's interests to be asserted interstitially. As a major shareholder in Mexican agricultural operations, he was best served by the construction of the dam with treaties to be worked out in the future. By arguing that debates over the building of the All American Canal (which would hurt his interests) should be postponed, it would appear that Chandler wished to move the debates into a less public forum where he would be better able to assert his power.

In a longer pamphlet, Chandler elaborates on his opposition, calling the pamphlet "The Other Side of the Question" (The Colorado River Land Company 1924). Chandler suggested that the purpose of the pamphlet was to "give the public exact information concerning the Colorado River problem, particularly with reference to the relation of the enterprises on the Mexican side of the boundary to the Imperial Valley" (*ibid.*: 3). Furthermore, the pamphlet claimed to

> expose oft-repeated falsehoods with reference to the contribution of farming interests in Mexico to the support of Imperial Irrigation District and to analyze the features which in our estimation have always made the Swing-Johnson plan unfeasible and impossible, though not objectionable so far as our selfish interests are concerned" (*ibid.*: 4).

Chandler claimed legitimacy for the pamphlet by saying that "there is nothing [in the pamphlet] except what has authoritative standing, such as sworn testimony before congressional committees in Washington, certified exhibits from the records of the Irrigation District, etc" (*ibid.*: 3). He claimed that he was opposing the dam on "principle"—his own "selfish interests" were not a factor in the pamphlet. Indeed, the reader is led to believe that Chandler would actually benefit from the dam but opposes it regardless.

"Mexican land owners," Chandler argues, "do not oppose the construction of any practical flood control dam. As a matter of fact, no one opposes it. It is stated on the highest authority that if the project were relieved of the All American Canal and the power features, there would be no delay in having provisions made for it by Congress" (*ibid.: 5*). For Chandler, delay was the enemy. He argued at length that continual delays put the various valleys downstream at risk. This was irresponsible on the part of the greedy farmers who favor the All American Canal and those who laud public power.

Chandler calls the Canal a "$4,000,000 Canard" and argues that "facts belie statements" by the Imperial Irrigation District and Mark Rose (*ibid.: 16*):

> A persistent campaign of misrepresentation has sought to make it appear that land owners and operators in Mexico have opposed the proposed All American Canal because in some way not indicated it would injure their interests—that it would deprive them either of water or service essential to the success of their enterprises. Of course it has not been possible for us to answer all of the slanders against us.

The arguments of the Boulder Dam and All American Canal boosters amounted to little more than "slander" and a "campaign of misrepresentation." Chandler reacted to these "libelous," "malicious," and "utterly uncalled-for" "endless falsehoods" with counterpoints that continually denied the assertion of his own interests. He claimed a complete lack of bias, and pushed for the construction of the flood control dam immediately.

It is significant that Chandler did not use the oppositional discourse of Muir or the Sierra Club that emphasized preservation; indeed, Chandler has a vested interest in the construction of the dam. His interests lay with the development of Los Angeles through hydropower, drinking water, and flood control. His lands in Mexico, however, would be threatened by the construction of an All American Canal, and he was insistent that the plan be dropped in favor of a simple flood control dam. This position, though cloaked in terms of objectivity, supported his best interests. Power would be developed in due course, and delivered to Los Angeles to spur economic development. His and other land would be protected from flooding, and competition from the Imperial Valley would be reduced. Chandler was thus reacting to the proponents of Boulder in a self-interested, modernist vein.

Chandler's newspaper the *Los Angeles Times* also engaged in political work against Boulder Dam. In conjunction with the private utility company Southern California Edison, the *L.A. Times,* sent letters and telegrams on behalf of various public groups in southern California—without the knowledge or consent of these groups. In a bizarre scheme, the letters were sent to the members of the Federal Power Commission praising their action on Boulder Dam (specifically their cautionary letter to Congress). These letters were then published in the *L.A. Times* on 4 April 1924 under the headline "Federal Commission is Praised On Power Stand." The *L.A. Examiner* (a Hearst newspaper and supporter of the Dam and Canal) exposed the fraud on 4 April 1924: "Boulder Dam Plot Revealed: Fullerton Chamber of Commerce Secretary Says *Times* Printed Fake Wire Showing Him as Foe to Project." Chandler's newspaper was clearly a propaganda organ.

Boulder Dam was not just opposed by Southern California agriculture barons and power interests. It was also opposed by Southerners, who feared the competition in cotton production. Representative from Alabama and friend of Phil Swing, John McDuffe, forwarded some letters to Swing that had been circulated in the South. The letters described the competition to Southern farmers that would result from the extension of irrigated lands below Boulder Dam. Swing was told that Southern members of Congress saw the Boulder Dam proposition as being "full of dynamite to any congressman from the South who might vote for it" (Moeller 1971: 102). Though the source of the pamphlets has never been conclusively identified, Swing felt that the source was associated with The Colorado River Land Company of Harry Chandler.[21] As Elizabeth Sanders (1999) has argued, Congress's reflection of geographic differences pitted northeastern capitalists against periphery and diverse area farmer interests. Given the ambivalence of the Republican Presidents, seen largely as the party of eastern capitalists, the lack of support from Southern Congress members could have killed the project. Recognizing this potential, Swing recruited Frank McIver, of the Imperial Irrigation District, to help produce letters to members of Congress that responded to Chandler's pamphlet (Greer 1927; Moeller 1971: 102). The distribution of Chandler's pamphlet, and Swing's response, shows what was at stake for the economic elites of Southern California.

National pro-business groups also opposed Boulder, such as the national Chamber of Commerce. The United States *Daily* reported on 16 January 1928 that the president of the Chamber of Commerce presented their perspective to Lewis Pierson, chair of the Committee on Irrigation and Reclamation in the U.S. House. The letter expressed support for the "great importance and the purposes of national character which it will serve" in terms of flood control, inter-state water allocation, and international relations with Mexico. The Chamber, however, took issue with the question of power production and distribution, a feature of the project that the Chamber preferred be left to private interests. "Any legislation which is enacted with respect to the Boulder Dam project should expressly and affirmatively provide that all proper effort shall be made to have private enterprise receive such opportunity to generate and distribute power at Boulder Dam." There was, then, some national level opposition that was not the product of Chandler's organizing, yet the opposition was of a piece with his concerns about private interests. These national concerns were occasionally expressed by the national press, such as the Chicago *Daily Tribune,* which reported a comment by Secretary of Interior Hubert Work on 3 May 1928 that "Boulder Dam an Engineering Sin." Later that summer, on 25 July 1928 the *Daily Tribune* wrote a short opinion piece on "The Boulder Dam Deception" which made similar arguments.

In June and July of 1924, new opposition groups became vocal about Boulder Dam politics, raising less overtly partisan issues. News articles regarding the formation of the Colorado River Control Club show that this new organization, formed out of Brawley, a town in the Imperial Valley, favored the dam for flood control, water storage, irrigation, and hydropower purposes. They opposed Phil Swing, however, and the Boulder Dam Project bill as it stood, but it was unclear as to why. At the same time (July 1924) a man named Ed Sample declared his candidacy for Swing's congressional seat, and this Colorado River Control Club backed him. Their opposition appears to be due to the "unfeasible" character of the All American Canal, specifically their idea that the canal would cost too much and, following Chandler's early charges, the canal had not been approved by any "reputable engineers" (as quoted in the Indio *Date Palm,* 4 July 1924). The Club argued that by deleting the canal authorization portion the bill, the legislation would pass more quickly, and thereby give them flood control and reduced risk more immediately—but would deny the IID irrigation water.

This argument is the foundation upon which Ed Sample, discussed previously on page 70, built his candidacy. Sample argued that Swing and Johnson could have brought flood control to the Colorado much sooner. One newspaper ran a cartoon featuring Phil Swing asleep at the switch, with Boulder Dam being held up because of it (Figure 4.12).

Although the contested election diverted much time, attention, and resources away from boosting the dam, Swing defeated Sample by a great margin. As with the letters circulated in the South regarding cotton production,

Figure 4.12 Undated and uncited (*New York Times*) political cartoon from the Phil Swing Papers, UCLA.

Swing attributed the candidacy of Sample to the Chandler coterie: An undated and uncited newspaper clipping in Swing's papers at UCLA reported that the entire scheme was an attempt to undermine support for the All American Canal by opponents of the dam who were in league with Chandler.[22]

Chandler's opposition to the particular form of Boulder Dam proposed by the Swing-Johnson bill thus contained many different elements and tactics, some more underhanded than others. Overwhelmingly, however, his opposition must be viewed as modernist, self-interested, and defensive. Chandler was forced into reacting to the proposed legislation in defense of his interests in Mexican agriculture and the private development of Los Angeles. This regionally based opposition was not alone in being threatened by the Swing-Johnson bill. National groups also saw their interests threatened by the legislation, and responded with similar vehemence as Chandler's.

The National Utility Association and the Water Resources Association

The National Utility Association (NUA) reacted with alarm to the proposals to build Boulder Dam. They were interested in building dams on the Colorado River, and they found it acceptable for the government to finance construction due to the major expenses involved. They were not, however, interested in having the government distribute the power produced at Boulder. The NUA fought the Boulder plans in Congress with tactics that were both overt and subtle.

In a flyer circulated to Utility Company executives, the NUA urged members to oppose the Swing-Johnson bill. They argued that the production of power at the dam will "no longer amortize its entire cost" due to the high cost of construction, the cost of transporting the power to distant markets, and the decrease in costs of producing steam power. Singling out the inclusion of the All American Canal in the bill, the authors stated that the possible income from sales of power could not repay the cost, making the entire project suspect. "It is apparent, therefore, that the substitution of Boulder Dam power for power generated by the modern steam plants would not result in a saving sufficient to pay the Federal government any price for Boulder Dam power generated at the plant" (National Utility Association, no publication date).

The authors made a similar argument regarding the provision of drinking water to Los Angeles: The cost far outweighed the benefits, they argued. Furthermore, the drinking water that the city required was currently being supplied by the Owens River. The additional amount from the Colorado would be enough for 7,500,000 inhabitants and it would take "a great many years" before "it can be expected that this additional number of inhabitants will make their abode in Southern California."

The authors concluded that:

> the control of the Colorado River should be considered by the Federal Government from the original basis of flood control, leaving the matter of power development on the River as a whole to be undertaken from time to time as economic justification therefore presents itself.

The authors of the National Utility Association flyer argued point by point against the boosters' attribution of benefits to Boulder Dam. Similarly to Chandler, the NUA questioned the utility of hydropower and drinking water for Los Angeles and southern California. Their rhetoric downplayed their own interests in private power development and instead emphasized the cost of the project. Similar discursive devices can be seen in the Reg Manning cartoon (Figure 4.13) of California depicted as a hog bathing in a trough of river water and cooled by the fan of free power. Such gluttony, the cartoon argues, should not be paid for by U.S. taxpayer.

The flyer put out by the NUA was directed primarily at utility company executives. This was not, however, their only effort. They also hired journalist Frank Bohn to edit two pamphlets that summarized the conflict and made recommendations regarding the most fitting solution. "The Boulder Canyon Dam: The Essence of the Swing-Johnson Bill" (Bohn 1927a) was a pamphlet that argued the NUA's position, but claimed to be an objective treatment of the conflict. In support of this claim, the pamphlet's cover noted that "Supporting evidence for the statements contained herein may be found in the official documents, engineer's reports, charts, maps and other authoritative and official information compiled in a more comprehensive form entitled 'Boulder Dam.'" This reference was also a pamphlet produced by the NUA: "Boulder Dam: From the Origin of the Idea to the Swing-Johnson Bill" (Bohn 1927b). This publication was indeed a collection of speeches, reports, and reprints of the Swing-Johnson bill and House of Representative Committee Hearings. The legitimacy of the publication hinged on the authority of these sources. The pamphlets used these sources to argue their case that the Boulder Dam bill was, at base, "a bill for the benefit of California" (Bohn 1927a: 26) which, in many ways, it was.

The first pamphlet reprinted many public documents and speeches. These were collated in such a way as to make the aim of the pamphlet clear: it claimed to be an exposé of "Efforts to Establish the Government in Industry" (Bohn 1927a: Pamphlet Cover). The pamphlet argued that electric power was the basis for the development of the industrial advances of the United States, and quotes speeches by the President of the American Federal of Labor and an official report of the Minister of Labor of the British

Figure 4.13 *The Arizona Gazette* 5 September 1925

Empire. The pamphlet also "frankly reveals" Senator George Norris as a supporter of "government ownership" of sources of electric power (1927: 15). The pamphlet continued by printing reports of huge losses by the Bureau of Reclamation (1927: 18), arguments in opposition to the All American Canal (1927: 40–45) and arguments about the Los Angeles water supply (1927: 46–51). The argument regarding state's rights, however, was given the most sustained and vicious treatment (1927: 24–39). The conclusion of the section is that there is "great menace to the western states" if the "power of Congress to override the state of Arizona or the state of Utah should be established" (1927: 25).

The rhetoric of the pamphlets published by the NUA was of a piece with the rhetoric of Harry Chandler. The pamphlets argued point by point in a rational style that relied on its sources for legitimacy. These arguments reacted to the proposed legislation in a modernist vein, and left the interests of the publishers in the background. The pamphlets remained, of course, self-interested.

The Water Resources Association (WRA) also opposed Boulder Dam in a defensive mode. The group wanted to protect private industry from any encroachments by the government. This group was an eastern association, based out of New York, but their distance from the immediate effects of the dam did not neutralize their critique. Their rhetoric was individualist and modernist: as with Chandler and the NUA, the WRA was interested in dominating nature and turning it to human ends—but they wanted individual capital to do the developing and reap the rewards. Going further than the NUA, however, the WRA wanted to construct a firewall between business and government, and they used red-baiting tactics to further their ends: "Government in business is socialism, and socialism is a present and sinister threat against the foundations of the United States." The authors also quoted former President Coolidge: "The alternative to private ownership and control is public ownership and control. Broadly extended, this is communism"(Water Resources 1928: 20–21).

The WRA also argued point-for-point with the supporters of the Swing-Johnson bill. They suggested that there had been "insufficient investigation and engineering studies," that there was "no immediate urge nor will there be for years to come" to increase the water supply for Los Angeles, and that flood control "can be done in a better way by at least one other place and in very much less time and cost" (Water Resources 1928: 1).

Such strong and inflammatory rhetoric fanned the flames of opposition nationally. This rhetoric, even though pitched as unbiased or objective, was firmly self-interested. Many other groups and individuals produced oppositional discourses on the same model. For example, Nelson, Cook and Company, of Baltimore, argued against government intrusion into industry (Nelson Cook and Company 1927). But none opposed the development of the river in the abstract. Their opposition was only to the aspects that threatened their interests, and their style of rhetoric construed opinion and political assertions as statements of fact. This modernist tactic of engaging in rational discourse—one that disguised self-interest—was not the exclusive domain of economic elites, however. The opposition to Boulder Dam was also marked by interstate rivalries that are best described as self-interested and defensive modernist discourse.

Interstate rivalries: Arizona v. California

Arizona had been unhappy about Boulder Dam since the six-state ratification of the Colorado River Compact in 1922. Officials from Arizona felt that the division of the waters was not equitable in that California took more than its share of the water. Arizona thus refused to ratify the Compact.

At the hearings before the House Committee on Irrigation and Arid Lands, regarding House Resolution 2903 for the development of the lower Colorado River Basin, March 25, 1924, Arizona Governor W. P. Hunt opposed the bill, explaining why the state did not sign the Colorado River Compact:

> The people of Arizona are patient, long-suffering, and tolerant, but are becoming weary of the imperialistic designs of some neighboring States. This State comprises 42 per cent of the drainage area of the Colorado River Basin. . . . California contributes only 6,000 square miles to the drainage of area of the Colorado River and undertakes to claim majority portion of its benefits. . . . [The] proposal of Secretary Work, of Colorado, and California officials to nationalize the river, as proposed in the Swing-Johnson bill, is insulting. Arizona will not ratify the Colorado River Compact, as it is a proposal wholly unfair and unjust to this State. The people of Arizona who hope to see the river developed and who are opposed to the Compact have been opposing it in a manner to create as little bitterness as possible.

The Governor argued that the Compact did not distribute water proportionately based on contributions to the River. California was not deserving of the amount allocated, and was thus imperialistic—a charge that is not inaccurate. As this chapter has argued, social groups in these states, as well as the federal government, were indeed engaged in state building that explicitly aimed at building an empire. As Hunt uses the term, however, Arizona is the colonized region rather than the beneficiary of federal largess.

Hunt did not argue against building dams, however. He opposed the division of the waters but he did not oppose the development of the river. Arizona officials, in fact, were quite interested in building dams on the Colorado River. The proposed Bridge Canyon Dam was rejected—in favor of Boulder—as being too small. A proposal for a dam at Glen Canyon Dam competed with Boulder for some years in the early 1920s, but would have to wait some 40 years to be realized (see chapter six). This dam would have been too far from Los Angeles for that city to be a market for power or irrigation water, thus cities in Arizona would have access to the hydropower and irrigation water provided by a development of the Colorado at Glen Canyon.

In a supporting role for Governor Hunt, Mr. H. B. Hovland, of Tucson, argued in the same hearings that a perfect solution to the interstate problems was for the people of Imperial Valley to simply move to Arizona:

> There is in the record testimony, in several hearings, to show that Imperial Valley is in a desperate situation. It is located under sea level, and it is subject to floods and dangerous menace. I raise the question before this committee as to whether this is not the time to look into the question in a broader way, as to whether it is good business to put in three or four or five hundred million dollars (as we will before the river is developed) under a continual, perpetual menace to the lands and the people below. Is it not time that we begin to think of clearing the river below all of those great quantities of water and dams, removing the people to other places and making it safe?

In an interesting reversal, Hovland used the Boulder boosters' flood menace argument to contest the dam. Hovland pointed out that by simply moving the people out of Imperial the same ends could be reached by spending much less money. Better yet, that money could be redirected to the upper Colorado River, to build Glen Canyon Dam, providing prime farmland for the migrants. Hovland's testimony shows the extent of the vitriol between California and Arizona.

Governor Hunt further outlined Arizona's interests in a roundtable of articles in *Community Builder,* a construction periodical. Governor Hunt was joined in the debate forum by William Mulholland of the Los Angeles Water Bureau, E. F. Scattergood, of the Los Angeles Bureau of Power and Light, and C. C. Young, Governor of California (1927). Hunt's article, "The Arizona Position" made it clear that "Arizona constitutes 43% of the entire drainage area of the Colorado River system" and that Arizona "is entitled to a fair share of the benefits which will come from harnessing the river" (1927:20). California, Hunt argued, "constitutes less than 2% of the drainage area, produces no water in the Colorado River system and has exhausted her own natural resources" (1927:21). This defensive position pits Arizona against the behemoth of the West, California—a state that at the time had more population, industry, and political clout than all of the other western states. Though Arizona's officials did not downplay their own interests as Chandler, the NUA, and the WRA did, they argued in much the same vein. Theirs is a defensive modernism that did not oppose development, but rather supported equitable progress that equalizes power between large and small—an argument that resonates clearly with Progressive-Era ideals.

To counter this uneven power, Hunt opened up the question of whether the dam should be developed by the federal government at all.

"Private enterprise is willing to undertake the development of the river, in which event the states in which the dams and power plants would be located would be able to derive taxes for state purposes," a state of affairs that would not hold for federal development (1927:22). Driving this wedge between the seven western states and the federal government, Arizona attempted to gain power vis-à-vis California to acquire a larger water allocation.

In the same collection of essays, Mulholland responded vehemently (1927: 24).

> The time for quibbling and bickering on the question of Colorado River control and development has passed. For years we have been listening patiently to the half-baked objections and alternative plans of persons and groups seeking to divert attention from the one and only practicable plan for the control and development of the Colorado.

Mulholland deprecated the opposing interests and counter-posed their "half baked" objections to the "practical" quality of the plan as it stood. E. F. Scattergood (1927) similarly argued for the Boulder Dam bill by suggesting that it must be justified by a cost-benefit analysis. He found that the plan made economic sense, with the added advantage that it would protect lives and property from flooding.

The responses to Arizona's opposition were uniform: Arizona was being irrational ("half-baked") while California and the other states were rationally agreeing to a clear cost-benefit analysis that put everyone ahead. This construction of two sides, one rational and one irrational, mirrors the very discursive construction of the nature-society relationship. While the river is constructed as a menace in need of harnessing by the rational science of society's engineers, Arizona is portrayed as irrational for insisting on a "fair share" of the water allotment while the other states are reasonable in their acceptance of an unequal split. Such argumentation shows the ways in which Progressivism exhibits, as McCormick called it, a baffling diversity. The proponents of Boulder Dam could argue in a Progressive spirit for the greatest good for the greatest number, citing efficiency and practicality, all the while decrying the irrationality of equity.

H. S. McClusky, a member of the Arizona Colorado River Commission, took on this portrayal of his state when he was invited to speak to the University Club Pasadena on 6 June 1927. McClusky argued the same points as Governor Hunt, asking that the people of California give Arizona only what it was "morally, ethically, and legally entitled." McClusky declared that the fight was one of "life or death," and the citizens of Arizona "intend to live if we can." In a final one-sentence paragraph of the speech, which in the notes was crossed-out by hand, McClusky said "Owens Valley, in your state, is now writing the history of how people may be expected to react when they believe they have been wronged."[23] The people of the Owens Valley, on the eastern escarpment of the Sierra Nevada, had responded violently to the construction of the 250 mile long Los Angeles Aqueduct that funneled water from the Owens River and Mono Lake to slake the thirst of Los Angeles. Starting in 1924 and continuing for three years, saboteurs used dynamite to destroy sections of the aqueduct, head gates, and to fill portions of the route with landslides (Walton 1992). Clearly McClusky believed that Arizonans were on the brink of such endeavors, but declined to present such animosity to the University Club of Pasadena.

The Arizona newspapers had a field-day with the California-Arizona feud. Popular conceptions of the situation portrayed California, and more specifically the Metropolitan Water District of Los Angeles County, as an octopus. This allusion to the monopoly power of the nineteenth century railroad barons was a striking reminder of the powerful political discourse of progressivism in the early twentieth century.

Arizona's opposition to Boulder Dam crested with the arrival of the final Swing-Johnson Bill in the U.S. Senate in 1928. The bill had passed the House on a voice vote earlier in the year, but when it came up in the Senate it was clear that there was going to be an all-out fight. The Senate version was immediately held up in filibuster by Arizona Senators Carl Hayden and Henry Ashurst. Johnson tried to extend the session to save his bill, but he failed to break the filibuster. As one historian has noted, "the Senate adjourned in wild disorder. There were threats of fist fights on the Senate floor, while onlookers shouted and hooted in the gallery. An executive session was called for the final hour, depriving the spectators of the scene of the 'wildest, most emotional Senate since war days'"(Moeller 1971: 115–16, the quote is from the Washington *Star*).

The bill was held up through the close of the session, but was immediately reintroduced at the beginning of the next session. Given the momentum from its passage in the House and the flagging sentiment for Arizona, the Senate voted for cloture, ending the filibuster and bringing the bill forward to a full vote of the Senate. On 14 December 1928 Arizona's last effort to kill the Swing-Johnson bill failed in the face of an overwhelming vote in the Senate, 64 to 11.

Opposing the Form of Boulder Dam

In a perspicacious editorial, The Huntington Beach *News,* on 4 July 1924 summed up the various controversies. The editors start by noting that there seems to be "no objection" to Boulder Dam. Instead, there is contestation over how the dam will be built and who shall benefit. The editors fully agree with the importance of Boulder, saying:

> The construction of this dam is one of the imperative necessities of this generation, and for that reason there is no doubt whatever about the ultimate completion of the vast project. The future of the Imperial Valley of California, one of the richest agricultural areas in the Nation, is wholly dependent upon some kind of flood control of the Colorado river. . . . The objection to the project comes from those who oppose the power possibilities of the Boulder Dam proposition.

Indeed, those elites who opposed Boulder Dam did so precisely because of the threat to their interests from the public distribution of power. The editors suggested that this distribution would, in fact, be beneficial to all seven of the Compact states. This opposition was, therefore, unfounded.

The editorial continued, supporting the idea of public power:

> Why should not the power possibilities be developed by the Government at the same time the dam is build, and not left for private companies to later develop for the benefit of their stockholders?

The editors came down on the side of the government and the public, against the private utilities. In short, the editors argued for a public-interest state project that would not benefit the private utilities. These utilities, they further argued, were the very ones who were objecting to Boulder Dam, and they did so in their own interest, not in the interest of the people. The objections, the editors continued,

> come from Americans who own large tracts of land in Mexico; and with the canal passing through Mexican soil, as it does now, permits these owners to purchase water at very advantageous rates from the Mexican officials, as when the water passes beyond the international boundary the Government of the United States loses entire control of it. Mr. Swing is to be commended for his efforts in behalf of this great project.

Without naming him, the editorial indicated that Harry Chandler was behind the objections to the dam. They pointed out his economic interests, as well as the interests of the Mexican officials. By exposing these interests, the editors could ethically support Phil Swing's position in favor of a dam in the public interest. They did not, it should be noted, make any mention of Swing's own interests in the Imperial Valley.

The defensive oppositional discourses at Boulder Dam thus contested the form that the dam would take, but not the dam itself. Harry Chandler, the state of Arizona, and the National Utility Association all had vested interests in making sure that Boulder Dam had a particular set of qualities. Chandler wanted to make sure that an All American Canal was not built, Arizona wanted more water, and the NUA opposed federal power production. All oppositional discourses at Boulder were thus defensive and self-interested in their opposition. No opposing groups used a rhetoric of natural value to contest Boulder Dam. John Muir was never mentioned, and any aesthetic concerns were expressed only over the design of the dam. All social groups seemed to concur that nature was an entity to be dominated, not preserved. The only question was: Who would benefit from the domination?

CONCLUSION: CHANGING A MENACE TO A RESOURCE

The discourse of imperialist modernism about Boulder Dam was a monolith. There was no effective opposition to Boulder Dam, there was only disagreement over the form that the dam would take and who would gain the benefits that the Dam produced. Boosters used an imperialistic modernist discourse that combined elements of domination: control of nature, control of the economy, and control of the population. The various arguments used to justify this control ranged from taming the floods, reclaiming land through irrigation, and producing hydropower. Combining these arguments, the boosters claimed to be building an empire in the West. The several strands of this imperialistic discourse all contributed to the discursive construction of the river as an economic resource that would promote economic and social progress in the West. The strands were drawn together to form a generalized argument of "national importance" that highlighted the imperialism of the discourse.

Behind this monolithic discourse of the national importance of Boulder Dam lay a discursive construction that turned the river from a natural menace to a natural resource. The discourse at Boulder Dam was univocal on this element, if not on any other. All groups discussing Boulder Dam agreed on the necessity of dominating nature and stopping the Colorado River. Some groups, such as Arizona elites, felt that a site up-river would be better. Others, such as Los Angeles elites, felt that Boulder should be developed by private interests rather than under the direction of the federal government. All agreed, however, that the Colorado's flood menace must be tamed. In this respect, some version of Boulder Dam was a juggernaut: Nature would be subdued on the Colorado.

Due in part to the power of the imperial modernist ideology of state building and of the control of nature, it was the federal government who undertook the massive job as a way to spur economic development, increase control of the region's economy and population, and to exert a definite control over the territory of the West. This job was done through discourse as much as it was done through the physical construction of the dam, for in order to legitimately build the dam, there had to be a common understanding of the proper relationship between nature and society. Harold Ickes used the term "cooperation" to describe this relationship of domination, though this cooperation clearly benefited human society rather than nature.

And so, as Pettitt wrote, Boulder Dam was built. The dam now stands as an icon in the West. It represents the power of humans in their ability to dominate nature, the stability of the United States in the face of economic depression, and the simplicity of form that defined the Modernist Age. It is because of this multi-vocality that Boulder has come to be one of the major tourist stops as travelers "do" the West. Indeed, the dam looms large in popular culture, continuing to fascinate artists, poets, and musicians through the end of the twentieth century. This resonance is in part due to Boulder's monumental status: in the words of Richard G. Wilson, Boulder Dam is the quintessential monument of the moment of Modernism he calls the "Machine Age."

Wilson's Machine Age refers to the time when the machine dominated "all areas of American life and culture" (Wilson 1986: 23). This was the period in American life when a new era dawned, and it was the same sensibility that underlay other forms of artistic expression, such as Art Deco, the Bauhaus, Futurism, and Cubism; but the American inflection was through the art of the machine—both in design, function, and output. Wilson argues that large multi-purpose dams such as Boulder and Grand Coulee, the subject of the next chapter, caught the imagination of Americans in the 1930's (Wilson 1985: 465). It is not by accident, for example, that *Life* launched its publication in 1936 with Margaret Bourke-White's photograph of Fort Peck Dam in Montana on the cover. Boulder Dam, for Wilson, plays a similar role in the development of the American machine aesthetic.

The idea of the machine aesthetic, which so captured the imagination of Americans in the 1930s, coincided with the golden age of dam-building in America. The ideology of the machine was transcendent, and with it the ideas of progress, efficiency, and the rational application of science and technology to problems both political and social. Boulder Dam exemplifies this ideology in a peerless manner. Only Grand Coulee comes close to rivaling Boulder's grandeur. Though they are of different designs (Boulder is an arch dam, Grand Coulee a gravity dam), the plans for both were made by engineer John L. Savage and architect Gordon B. Kaufmann, and the dams were both built by the engineering powerhouse Six Companies.

Henry J. Kaiser, a main partner of the Six Companies, was involved with most of the major construction projects on the West Coast in the first half of the twentieth century. He helped build Boulder, Grand Coulee, and Shasta Dams, and the San Francisco Bay Bridge. He built Liberty Ships for World War II, and was even considered as a running mate by Roosevelt in 1944. Kaiser was once asked to explain his success in the construction industry during the Depression. He responded with an anecdote: "Contractors are all alike. They start out broke, with a wheelbarrow and a piece of hose. Then suddenly they are in the money. Everything's fine. Ten years later, many are back where they started from—with one wheelbarrow, a piece of hose, and broke. So, before you work yourself out of the last job, line up a bigger one to pull yourself out" (Heiner 1989: 59). And this is exactly what Kaiser and the Six Companies did: they lined up Grand Coulee before they finished Boulder Dam.

Grand Coulee Dam poses an interesting case of the domination of nature and the character of state building in the American West. Boulder Dam was largely the product of a small group of farmers in the Imperial Valley, in league with the Bureau of Reclamation and a wide array of voluntary associations. These groups fought with private U.S. farming interests, with land in Mexico, and power companies in the Southwest. In the case of Grand Coulee the opposition was similarly from a defensive, self-interested, and modernist position of local economic elites. But the governmental interest in Grand Coulee was far less forceful, and the role of groups and individuals in civil society was much more important to the building of the dam. Though construction was begun while Boulder was still being completed, allowing Kaiser to "pull himself out" of finishing the Boulder Dam job, the contrasts are as important as the similarities.

Chapter Five

Grand Coulee: "Mightiest Thing Ever Built by Man"

Roll on, Columbia, roll on,
Roll on, Columbia, roll on.
Your power is turning the darkness to dawn
Roll on, Columbia, roll on.

Tom Jefferson's vision would not let him rest.
An empire he saw, the Pacific Northwest.
Sent Lewis and Clark, and they did the rest.
It's roll on, Columbia, roll on.

And far up the river is Grand Coulee Dam,
The mightiest thing ever built by man,
To run the great factories and water the land,
It's roll on, Columbia, roll on.

"Roll on Columbia," Woody Guthrie

In the misty crystal glitter
  Of the wild and windward spray;
Men have fought the pounding waters
  And met a watery grave.
Well, she tore their boats to splinters
  But she gave men dreams to dream
Of the day the Coulee Dam would cross
  That wild and wasted stream.

"Grand Coulee Dam," Woody Guthrie

Within the discourse around the proposal and construction of Grand Coulee Dam, we can hear the echoes of debates over Boulder. Built just after Boulder Dam (1931–36), Grand Coulee (1933–41) benefited from public confidence in such projects, a positive governmental climate toward dam building, and technological achievements invented at Boulder. Built on the successful legal foundations of the Colorado River Compact and other enabling decisions, Grand Coulee was also completed by some of the same construction companies and many of the same workers who built Boulder. The continuities are certainly strong, but the contrasts are also important: local boosters were key to the success of Grand Coulee Dam. This chapter explores the discourses and rhetorical strategies of the various players, both local and national, in their "23-year battle for Grand Coulee Dam" (Woods 1944). These opposing camps are analyzed in terms of the competing and overlapping discourses that they used. These discourses are all built on particular constructions of nature-society relationships that enabled human domination of the Columbia River, in Guthrie's lyrics "that wild and wasted stream." These discourses about nature were based in an ideology that helped construct the river as something to be dominated by humans. The river was seen as a wild entity, as Guthrie calls it, a river of "wild and windward spray;" but one that could nonetheless be harnessed by human endeavor. High modernist discourses characterized much of these rhetorical styles used at Grand Coulee. The dam was part of an imperialist vision and was to be built by the federal government as part of a plan to settle and build up the West. These typical state building goals were implemented using the scientific and rational engineering techniques that would carry the region, and therefore the nation, along the linear path of progress. As Guthrie labels it in his lyrics to "Roll on Columbia," Grand Coulee Dam is "The mightiest thing ever built by man." Ironically it stops the river from rolling.

THE SOCIAL, POLITICAL, AND GEOGRAPHIC HISTORY OF GRAND COULEE DAM

Built in the late 1930's, Grand Coulee Dam was a technological achievement that rivaled Boulder Dam in its size and difficulty of construction. Though Boulder Dam is higher—726 feet from bedrock, where Grand Coulee Dam is 550 feet—Grand Coulee is wider and contains more concrete—four fifths of a mile at the crest of the dam and 10.5 million cubic yards to Boulder's quarter mile crest height and 3.25 million cubic yards. These differences are largely due to the fact that they are different types of dams. Boulder Dam is a classic arch dam that distributes the force of the water across the dam and onto the walls of the canyon. Grand Coulee is a gravity dam, where the force of the water is held back simply with the mass of the concrete. The differences in construction are dictated by the geography of the site: the Columbia River flows through a much wider canyon than the Colorado at the point of the dam. The Columbia River also carries more water per year than any other river in North America except the Mississippi: 160,000,000 acre-feet (a-f)[1] per year (ten times the flow of the Colorado River). Its tributaries draw from Oregon, Washington, Idaho, Montana, and British Columbia (Lang 1992; Williams 1951). The history of Anglo settlement on the Columbia is a history of human struggle with the forces of nature and a negotiation among inhabitants for political, economic, and social control of the region and its main resource, the river.

Following the "discovery" of the Columbia River, the United States and Britain disputed ownership of the Pacific Northwest region. John Jacob Astor's fur company was established at the mouth of the Columbia just four years after Lewis and Clark began the mapping of the region in 1806. Just a few years later, as the English lost the War of 1812 and focused entirely on their Canadian holdings, the Pacific Northwest region became the last frontier of conflict between England and the U.S. The region was held jointly until 1825, but soon thereafter Americans flooded the area with immigrants following the Oregon Trial. The English tried to hold on to their lands north of the Columbia, using the river as a dividing line, but the Americans held firm at the latitude of 54 degrees 40 minutes, the line of lowest Russian settlement. President Polk was said to have been elected in 1844 in part due to his hard-line stance on "54–40 or fight" (Williams 1951).

By the early twentieth century, the Americans had built a successful local economy largely based on agriculture, fishing, and fur trading (White 1995). Many areas of the coastal Pacific Northwest receive a large amount of precipitation, so the development of the Columbia River was pushed by slightly different factors than the development of the Colorado River. Up until the early 1900s, private and local attempts had been proposed to control the river, but all were too expensive for private capital to undertake—a familiar story in the West. It would require the federal government's deep pockets to control the river.

In 1925, the United States House of Representatives ordered the Army Corps of Engineers (ACOE) to study the nation's rivers looking for navigation, flood control, irrigation, and hydropower locations. The report recommended a series of ten dams on the Columbia River, the first being at Bonneville, another at Grand Coulee.

Grand Coulee Dam added irrigation to the list of development purposes on the Columbia (previously navigation and flood control had been primary). The coulee (a geological term for a deep gorge) was a product of the last ice age, where a lobe of the glacier dammed the river and forced it to cut through an ancient lava bed, opening a gorge—the coulee. When the glacier melted away, the river assumed its old course, and left the Grand

Coulee open. The idea of damming the coulee is credited to William Clapp, a local Washington state politician, who pursued the idea for 15 years, starting in 1918 (Ficken 1995). The ACOE, however, declined to pick up the project because the proposed dam was too big. It would have to be over 4,000 feet wide to stretch across the river, and hundreds of feet high to be within reach of pumps to deliver irrigation water to the farmland surrounding the coulee (Reisner 1986).

Though the ACOE declared the project technically possible in 1931, it proposed a 200-foot high dam, as it had built at Bonneville. Such a dam would produce power, improve navigation, and control flooding—these elements being the main concerns of the ACOE. But such a dam would not do for irrigation. To supply water to surrounding farmland the spot needed a high dam because the canyon rim would be another 500 feet from the water level of a 200-foot high dam. Not only did the pump technology not exist at the time, but the pumping distance would also require too much power. A high dam, on the other hand, would decrease the pumping distance by raising the reservoir level and produce enough extra power to divert some to the pumps and distribute the rest across the region. This would allow the rich soils surrounding the coulee to be farmed. By the time Franklin D. Roosevelt adopted the idea of Grand Coulee, moving people onto good farmland was a strong political idea. Roosevelt supported the building of Grand Coulee as a New Deal project in part for the jobs that the construction project would create and in part for the farm lands that it would bring into production (Pitzer 1994).

Congress balked at the second dam on the Columbia (Bonneville being the first), and would not approve the funds. It would require $270 million in 1933 dollars to build Grand Coulee. Congress, ever concerned about "pork," would not authorize that amount of money to a sparsely inhabited corner of the West. Congress had, however, granted Roosevelt an essentially open checkbook and widespread power in the Public Works Administration (PWA), part of the National Industrial Recovery Act, passed in June of 1933. Roosevelt authorized Grand Coulee with $63 million in seed money from PWA emergency funds, against the wishes of many parties. Roosevelt faced resistance from the ACOE due to technical barriers, and so he turned to the Bureau of Reclamation in the Department of the Interior. Here, however, there was also dissent from his own Secretary of the Interior, Harold Ickes, who argued that the nation could only afford one project at a time, and that one should be Bonneville (Sundborg 1954). Roosevelt was not swayed and he persuaded Congress to add some money to the PWA funds, and ordered the project to move forward. The foundation for the ostensibly low dam was begun, and the first appropriation was spent by 1935. Roosevelt and the Bureau went back to Congress and asked for more, but this time they asked for enough money to build a high dam. The Bureau engineers revealed that the foundation they had poured could support a high dam.

The difference in the two types of foundations is drastic; a low foundation needs to support far less concrete above it, and does not need to be so large as to hold back a hundred-mile long reservoir. A high dam, on the other hand, requires a thick foundation—the foundation of Boulder Dam, for instance, is almost as long as it is high. To put a low dam on the foundation for a high dam would, according to one observer, be like putting the frame of a passenger car on the chassis and drive train of a Mack truck (Reisner 1986: 157). In the political climate of the New Deal, Congress was weak relative to the President; it did not dare revoke funding. Appropriations for a high dam were thus authorized. Roosevelt and the Bureau of Reclamation had managed to manipulate Congress and were successful in building the high dam that would support the multiple aims of the New Deal: jobs, hydropower, and new farmland (Reisner 1986: 157).

Furthermore, the initiative for this second grand public works project in the Pacific Northwest has been questioned by commentators (Reisner 1986: 158). The Pacific Northwest had about three million inhabitants at the time, and it was expected that the power produced from Bonneville itself would be superfluous—Grand Coulee would more than double this available amount. In addition, there was a great deal of agricultural produce going to waste already: bringing new farmland into production would hardly help this overproduction crisis.

Roosevelt, however, saw increased farmland as an opportunity, not as a liability. Grand Coulee was surrounded by prime farmland but the lack of water kept the land from being used. Concerned with migration and economic depression, Roosevelt and the Bureau of Reclamation acted in consort to bring the land around Grand Coulee into cultivation in order to move farmers from sub-marginal land elsewhere in the nation onto prime agricultural land in Washington State (Pitzer 1994). This, Roosevelt argued, would allow production where it could be optimized rather than where it contributed to ecological damage.

According to Reisner, almost everything involved with Grand Coulee was the largest in the world: The mass of the dam (10.5 million cubic yards), its crest length of 4/5 of a mile, the concrete-mixing plant, spillway, coffer dams, generators, and powerhouse. The reservoir that Grand Coulee Dam backs up is called Franklin Delano Roosevelt Lake, and it stretches 150 miles north to the Canadian border. The project was planned so as to irrigate

1 million acres of land, and provide massive amounts of hydropower. The construction process, however, was far from perfect.

Nothing like Boulder and Grand Coulee had been done before, so the lessons that were learned building Boulder were applied to Grand Coulee. This included such tricks as cooling the concrete by running water pipe through it. Different problems were faced in the Pacific Northwest, however. While Boulder Canyon faced tremendous heat, the Grand Coulee faced torrential rains. At one point, one wall of the gorge threatened to slide down and cover the foundation of the dam. The ingenious engineers assembled the largest collection of refrigeration units ever in one place and froze the slide. When the coffer dams (temporary dams that hold the river back so that the foundation for the high dam could be laid) sprang a leak, the managers plugged the holes with old mattresses (Reisner 1986:160).

In September 1941 Grand Coulee came on line, and its power system was connected to Bonneville and put under authority of the Bonneville Power Administration (a government corporation inside the Department of the Interior which, to this day, sells the power generated by the dams to private and public utilities and interests). The massive amounts of power that the two dams brought to the Pacific Northwest turned out to be critical to the outcome of World War II.[2] By the end of 1941, 92% of the total power from Bonneville and Grand Coulee was going to aluminum production. This helped produce the 60,000-odd planes that flew over Europe and the Pacific and helped win the war (Reisner 1986: 162). By 1945 Grand Coulee was generating 2,138,000 Kilowatts of electricity, the largest single energy producer in the U.S. Much of this surplus energy from additional reactors installed after 1941 was sent down river to the Hanford Nuclear Reservation via 20,000KW power lines, where Plutonium and Uranium was enriched for the Manhattan Project. Hanford produced millions of gallons of radioactive waste, much of which was unintentionally released into the Columbia River. But incredibly, researchers also *purposely* released radioactivity from the site to study its patterns of travel, exposing tens of thousands of citizens (Josephson 2002: 60–61).

DISCOURSES AT GRAND COULEE

While Boulder was proposed and supervised by the Bureau of Reclamation, Grand Coulee was initially proposed by local boosters. The discursive tactics were thus similar in some ways and yet they differed in important aspects. As with Boulder Dam, discursive strategies such as national-level imperialistic exhortations about manifest destiny were used. As exemplified by Woody Guthrie's lyrics to "Roll on Columbia," Grand Coulee Dam was portrayed as the fulfillment of a historic vision of empire in America. Grand Coulee could run a thousand factories for Uncle Sam, bring darkness to dawn; it is 'the mightiest thing ever built by man.' But another style of discourse was used, one that I term, following Weber (1952), Hebraic discourse about local chosen peoples in the promised land of north central Washington. Guthrie's lyrics to "Grand Coulee Dam" describe the brave pioneers who "fought the pounding waters" of the river. But the river gave these courageous men "dreams to dream" in a harsh frontier land. This land was described as both a "last frontier" and a "promised land."

But the discourse around Grand Coulee is more complex than just two competitors (see table 5.1). Local boosters used both imperialistic and Hebraic discourses in a high modernist vein to argue for the federal government to build the dam. In contrast, the national-level discussions were centered on an imperialistic discourse that was divorced from a Hebraic discourse. Instead, many in the national press and the Federal government actually viewed the local inhabitants as backward rural folks. They thus mobilized an anti-Hebraic rhetoric within the imperialistic discourse that disparaged the locals and envisioned a dam that would bring prosperity, and therefore a civilizing influence, to the region.

The imperialistic discourse was thus not a monolithic category. The two opposing local booster groups argued over whether the dam would be a high dam (dubbed the pumping plan) or a low dam (called the gravity plan). The high dam would harness the hydroelectric potential of the river in addition to the irrigation possibilities, while the low dam would allow only for hydopower and flood control. The low dam contingent was primarily made up of regional power industry elites who feared a high dam would undercut their prices and compete with their monopoly on power in the area. Their imperialistic rhetoric is thus suffused with an individualist capitalist orientation that values unfettered competition, and is suspicious of state-sponsored programs. The proponents of the high dam, on the other hand, argue from a high-modernist perspective that values the interventions of the state in building large-scale water systems that could not be achieved by local capital alone.

As with Boulder, these contesting discourses at Grand Coulee were thus over what *form* the dam would take, not whether there *should be* a dam. This exclusion of possibilities is important, for the success of Grand Coulee represents the victory of a state-centered high-modernist proposal over competing individualist and private industry-oriented proposals that would in all other respects similarly dominate nature. The striking absence of oppositional discourse shows the success of the high-modernist ideology in the American West in general and the Pacific Northwest in particular. As with Boulder, the

Table 5.1 Grand Coulee Discourses and Social Groups

	Discourse		
	Imperial High Modernist	Imperial Capitalist	Hebraic
Social Group			
Local Boosters	X		X
Local Elites		X	X
Local Press	X		X
National Press	X		X
Government	X		

opponents thus competed over details while agreeing on the building of some sort of dam. This situation represents a differentiation within the ideology of human domination of nature. The two competing dam proposals represented two ways of applying this ideology to the natural world; the individualist orientation failed in the face of the high-modernist, state-centered proposal.

The social groups that favored the dam can be broken down into two clearly opposing camps. On one side were the local boosters that backed the "pumping plan." This proposal called for a high dam that would combine flood control, navigation, irrigation, and power production purposes. Opposed to these boosters was an alliance of regional economic elites who favored a "gravity plan" that would eliminate the power production aspects at Grand Coulee. This debate exploded in the 1920s with furious local, regional, and national debates over public versus private power. The pumping plan advocates favored the high dam's cheap hydropower, which would produce more irrigation possibilities. The gravity plan advocates—made up largely of local elites connected in various ways to the private power industry—opposed such a plan for fear that it would compete directly with their economic interests.[3]

The major players in the Grand Coulee drama are thus local boosters, regional elites, federal agencies, and elected federal officials. On the pumping plan side the key individuals were Rufus Woods (a newspaper publisher in Wenatchee, Washington), James O'Sullivan (a lawyer and schoolteacher in Ephrata, Washington), and William Clapp (a lawyer in Ephrata, Washington). On the gravity plan side were J. D. Ross (superintendent of Seattle City Light company), executives of the Washington Water Power Company of Spokane, and other regional politicians, especially the Spokane

Chamber of Commerce, who sided with the power companies. The U.S. Army Corps of Engineers and the Bureau of Reclamation played an interesting role in between these two groups. The federal agencies were lobbied by both groups and at various times aligned themselves with one side or another. Federal officials such as President Franklin D. Roosevelt and his Secretary of the Interior Harold Ickes entered the fray at key decision-making moments.

GRAND BEGINNINGS: THE PUMPING PLAN AND THE GRAVITY PLAN

Paul Pitzer, in his history of the Grand Coulee project (1994), suggests that the actual facts regarding the origins of Grand Coulee Dam have become muddled given the subsequent rancorous debates. Both sides, he charges, may have exaggerated their claims and the timing of the original ideas. Determining exact timing in the historical record is often problematic; what is clear in this case is that two ideas for irrigating the Grand Coulee area were born near the end of World War I, and the debate over the dam grew into a major regional issue over the next fifteen years.

On 18 July 1918, Rufus Woods printed an article on page seven of his newspaper, *The Wenatchee Daily World*. The article described an interview with a local lawyer from Ephrata, a small town near Wenatchee, Washington. William M. ("Billy") Clapp suggested to Woods that the Grand Coulee could be used as part of an ambitious irrigation project. Woods published the article in his characteristically effusive phrasing and with the headline in all capital letters, as was his custom when he wanted to highlight an issue. It was entitled "FORMULATE BRAND NEW IDEA FOR IRRIGATION GRANT, ADAMS, FRANKLIN COUNTIES, COVERING MILLION ACRES OR MORE" (Woods 1918). Woods often capitalized certain words or phrases as a way of emphasizing them and forcing the reader to pay attention. In this case, he capitalized the entire headline, almost making it scream its importance to the newspaper-reading public.

The project that started as a search for news may have been a joke at first. At the moment of Woods' printing, however, the idea of a dam at Grand Coulee was not new. In fact, using the coulee as a storage basin for some sort of irrigation project was proposed as far back as 1892 (Pitzer 1994). When Woods rediscovered and published it, the idea had languished for some twenty-five years. Looking back twelve years later, in a 1930 letter, Woods suggested that "it was a joke to a certain extent but it began to 'take' all over this territory" (Pitzer 1994:17).

Woods and Clapp received a great deal of local attention due to the article, but the idea went nowhere for several years. It would take many more articles in the *Daily World,* and the efforts of several other local boosters before the idea became a viable plan. It was dubbed the "pumping plan" because it called for a dam at Grand Coulee and massive pumps, fueled by hydropower produced at the dam, to lift water up to the lands surrounding the gorge.

At nearly the same time as the pumping plan was making headlines in Wenatchee, Elbert F. Blaine proposed a plan to bring water to the area from Idaho. Blaine often traveled across the eastern Washington region for his employment with the state Railroad Commission. Blaine was a landowner and real estate developer in the Yakima Valley of Washington, and saw the potential for an irrigation project in the area. He surveyed the land between the Pend Oreille River in Idaho and the Big Bend area of Washington (a local appellation for the Grand Coulee area due to the nearly 90° turn that the Columbia River takes) and found that most of the distance was downhill. This would enable a gravity flow irrigation plan that could deliver water to the fertile valleys of north central Washington. Blaine proposed his plan in 1918 to the Spokane Chamber of Commerce, and the "gravity plan" was born.

IMPERIALISTIC DISCOURSES: IRRIGATED EMPIRE IN THE PACIFIC NORTHWEST

The fifteen years after World War I saw a great deal of contentious debate between the two plans. Both sides used a style of discourse that includes what I call imperialistic or empire-building rhetoric. This rhetoric, however, was differentiated by either a high modernist or an individualistic capitalist inflection. The imperialistic rhetorical strategy argued that the dam proposal would improve the economic might of the region. Examples of this rhetoric can be seen in the headlines from various newspaper articles. Some authors argued that the power would help the nation extend its empire into the Pacific Northwest: "Lower Cost Of Power To Help Nation" (Dill 1934). Others viewed the dam as a monument to American achievement: "'Eighth Wonder of the World,' Harnessing Mighty Columbia Steadily Rounding Into Form" (Harvison 1936). Still others saw the dam as costly, but worth the price due to its grandeur, placing America in the ranks of great civilizations: "Coulee: Gigantic Structure Costs Millions But Benefits That Will Accrue Amply Justify Definition–'Man's Mightiest Work'" (Fussell 1938).

The imperialistic discourse is based in the Progressive-Era ideology of unilinear progress grounded in a rational application of science and technology. The ultimate aim of those who used this discourse was the domination of nature and the use of natural resources for the benefit of society. Concern for nature was not expressed; on the contrary, nature was seen primarily as a force to be subdued (the "wild" stream). Such views betrayed the belief, on the part of local boosters, that the project was worthy of being built due to this challenge that nature posed. In the case of the Grand Coulee project the goal was the irrigation of regional farmland that would spur the local economy and increase population by making the area more attractive to settlers. These two processes of agricultural and population growth, the argument went, would help boost an industrial economy in the region. This two-sector economic development model was still new in the 1920s but it would quickly take hold in western capitalist economies and become dominant by the 1950s. This imperialistic discourse is the closest form of discourse to Scott's (1998) definition of high modernism, although it shows the complexity of how high modernism played out in the American West. In Scott's model of high modernism, it is the centralized and powerful federal government that pushes development plans on the locals. In north central Washington, boosters of the two irrigation plans used the rhetoric of economic growth to convince locals, state officials, and federal bureaucrats that millions of dollars should be spent on this economic model. Such economic growth would provide the foundation for an empire in the Pacific Northwest.

Rufus Woods was by far the most visible user of such rhetorical techniques. His newspaper published story after story boosting the dam under a masthead that declared the paper was "PUBLISHED IN THE APPLE CAPITAL OF THE WORLD AND IN THE BUCKLE OF THE POWER BELT OF THE GREAT NORTHWEST" (Woods 1937b). Wenatchee is the largest town in an area of north central Washington that was (and still is) heavily agricultural, the primary crop in the area being apples. Wenatchee had proclaimed itself the apple capital of the world, but Woods added the idea of the "buckle of the power belt" when the Grand Coulee project was approved. This phrasing is an attempt to rename the region with a colloquial phrase that evokes the idea of other regional industrial or agricultural foci, such as the Corn Belt in the Midwest. This turn of phrase, however, allows Woods to underline the agricultural and industrial combination in the Pacific Northwest that makes it a key to the growth of the American empire.

Under this masthead in the 16 December 1920 issue Woods called for a building of a great "Inland Empire" in which irrigation and power production went hand-in-hand:

> SUCH A POWER . . . WOULD OPERATE RAILROADS, FACTORIES, MINES, IRRIGATION PUMPS, FURNISHING HEAT AND LIGHT IN SUCH MEASURE THAT ALL IN ALL IT WOULD BE THE MOST UNIQUE, THE MOST INTERESTING, AND THE MOST REMARKABLE DEVELOPMENT . . . IN THE AGE OF INDUSTRIAL AND SCIENTIFIC MIRACLES.

In his all-capital paragraph (which must have required a special collection of tools) Woods used superlatives in describing the "Inland Empire" that Grand Coulee Dam would enable. The dam would improve the entire country by virtue of its capacity to help the Pacific Northwest make use of its rich natural resources through mining and irrigation. The dam would also provide surplus power, driving industrial production. This combination, Woods declared, would be more remarkable than any other in an "age of miracles." Though his rhetoric was often overblown, Woods' characterization was remarkably accurate: Grand Coulee would indeed be labeled, along with Boulder, as an icon of the Machine Age (Wilson 1986).

Furthermore, in a special edition of 28 April 1937, printed for the Apple Blossom Festival in Wenatchee, Woods outlined the special social and geographic characteristics of the region. Woods suggested that the north central Washington area was a "country of amazing resources" that contains the "greatest power stream in the civilized world." No other area has the "PROVEN POSSIBILITIES in the way of soil, climate, mineral resources and everything to make up an ideal habitat for man" and yet the inhabitants of the region needed the help of the government to build the Grand Coulee Dam. Within this high modernist discourse, the dam "is the key to future development of the Northwest," a future that would benefit the entire nation with its productive power. Woods boosted the dam with an understanding of its contributions to progress, efficiency, and the building of the "Inland Empire" of north central Washington.

In Woods' account nature played an interesting role. While in many imperialistic discourses nature was simply that which was to be dominated, to Woods nature was a willing participant. Nature had "built [a] 28-mile main canal" and had even provided a reservoir for use in the project. This complication in the views of nature is in part due to Woods' search for any and all points in favor of the project, and in part due to his relationship to the land as a local inhabitant.

In 1937, Rufus Woods expanded his influence by starring in a national radio program. WABC (New York) and the CBS networks produced a program, funded by DuPont, called "The Cavalcade of America." Session number 86 was entitled "The Eighth Wonder of the World," and featured Rufus Woods as the hero (Woods 1937a). The argument about why a dam could and should be built on the Columbia River was dramatized by the radio program using the model of an ancient Greek Drama, complete with chorus. Woods was cast in a David versus Goliath situation, fighting against a community representative (Abe) who believes the plan to be impossible. In the end, Woods convinces the community that it can be done:

ABE: You can't harness the river. It wastes time an' money.

RUFUS: A dam in the coulee will bring life to thousands of thirsty acres. Fertile fields will be born! There'll be thousands of fine homes where now there's nothing but sagebrush.

ABE: Crazy. Out of his senses.

RUFUS: We mustn't lose faith when we've really started! It's time to fight all the harder, not give ground but move forward! What do you say neighbors?

Woods had several allies in his boosting of the pumping plan; many people helped him publicize the plan, including groups such as Teepee Information Services. Teepee was comprised of local boosters who published various pamphlets and information sheets. They maintained an information booth and encouraged tourists to visit the dam site. The Teepee Information Services used a great deal of imperialistic rhetoric: the pamphlets they published are loaded with examples of grand claims regarding the region's potential economic power, natural resources, historical sites, and strategic importance to the nation.

Representative of such efforts, the "Curb-Side Story of Grand Coulee Dam and a Last Frontier" tells of local tourist sites, such as Chief Joseph's resting place, calling him "the Napoleon of his race" (Weil 1938). The pamphlet goes on to describe north central Washington as the last frontier, with its "rugged and untrammeled beauty" that is "yours to explore." Grand Coulee Dam becomes the "eighth wonder of the world," and the area is boosted as a place for industry to settle, as "noted economists urge decentralization . . . few localities may equal and none may excel the opportunity so afforded at Grand Coulee." The pamphlet sings the praises of the area, with its "slumbering riches," "tall, tall timber," minerals, agriculture, and stock-raising. Furthermore, the area is important for national defense—the area lies inland, with mountain barriers, and as such, it is the ideal spot for "a maximum of security for munitions plants and munitions storage depots. And why not a plant for the manufacture of airplanes?" (Weil 1938). The imperialism is clear in this last sentence. National defense would come to be crucial on the West Coast as America began to be concerned with the war in Europe and the military build-up in Asia. Indeed, this concern would become central to the success of the boosters of Grand Coulee Dam in their appeal for public approval.

As far as individuals are concerned, James O'Sullivan was Woods' most energetic ally. In fact, historians of Grand Coulee debate the relative importance of the two men (Ficken 1995; Sundborg 1954). Some authors have honored Woods for his publicity work, and point to this as the crucial factor; others suggest that the dam would never have been built without O'Sullivan's tireless lobbying in both Washington State and Washington, D.C. As with so many historical debates, the truth is somewhere in the middle. While keeping the project alive in the public mind through continual newspaper articles was certainly important, it was equally important to force politicians and government bureaucrats to take the project seriously. This last job of lobbying Congress and the various executive branch agencies was taken on by James O'Sullivan, who used a high modernist imperialistic rhetoric to convince the federal government that it was its duty to build up an empire in the Pacific Northwest.

O'Sullivan was a small-town lawyer and schoolteacher who devoted the better part of his adult life to getting the dam built. While Woods had a newspaper to run, O'Sullivan went into debt to work solely on boosting the dam. In 1929, a number of local boosters from the Wenatchee-Ephrata area met to form the Columbia River Development League. They selected O'Sullivan as Secretary, the only position that carried a salary, though meager. This was O'Sullivan's first official position as a Grand Coulee booster, but he had devoted a great deal of effort to building the dam in the preceding years. Through those years he and his family endured a voluntary poverty.

As well as traveling to Washington, D.C. to lobby for the dam, O'Sullivan wrote many letters to politicians attempting to secure funding for studies of the dam site, to keep the idea of the dam alive in the Bureau of Reclamation, and to argue the merits of the dam to congressional and administration detractors.

When Arthur M. Hyde, the Secretary of Agriculture, came out in opposition to the Columbia Basin project in March of 1932, O'Sullivan responded with a 61 page statement to the Secretaries of Agriculture and Interior. Hyde had opposed the project on the basis of agricultural competition. It did not make sense, Hyde argued, to bring new lands into cultivation when other lands in the South and East were suffering from overproduction problems. In fact, Secretary Hyde argued, no new lands would need to be brought into production until 1960.

O'Sullivan's response is a long diatribe, arguing with Hyde point by point. On each issue, O'Sullivan draws on the ideas of growth and progress that are central to the imperialistic discourse, while also showing the dire consequences of not building Grand Coulee, using words and phrases that connote downfall, backsliding, and eventual doom (O'Sullivan 1932: 2):

> If the secretary's estimates are correct and his reasoning sound, we face a very gloomy prospect in the United States and particularly in the West. We should retrogress not only in agriculture but also in industry and in every other branch of our activities that produce an apparent surplus. We should progressively retreat from our proud position as the greatest Nation in the world to eventual extinction. We should sit idly by while other nations are rapidly expanding the area of their agricultural lands . . . This is not the voice of statesmanship nor the voice that brought about the building of this great Commonwealth; it is the cry of despair, of defeat and annihilation.

O'Sullivan concludes with the connection of the West to the nation as a whole: "The statement by the Secretary attacks not only the Columbia Basin, but all reclamation [projects]. It seeks to undermine the future growth and prosperity of the West. All true friends of the West should rally in defense of the basin project . . . The law of life is the law of growth. We must now concentrate on building up our home markets if we are to prosper in the future. In the construction of the Columbia Basin project and other meritorious ones we can best serve man's end. The Basin project is needed today to provide for the future needs of the Northwest and the Nation" (O'Sullivan 1932: 48).

In a hearing before the House Committee on Irrigation and Reclamation in May of 1932 (72nd Congress), O'Sullivan testified regarding the Columbia Basin project. He was in good company, for Elwood Mead, Commissioner of Reclamation, and Major John S. Butler, author of the "308 Report" that recommended extensive development of the Columbia River, also testified at the hearings.

O'Sullivan made an impassioned argument for approval of the measure to fund the Columbia Basin project: "The 11 Far Western States can not make any further progress without reclamation. . . . If you stop reclamation in those states, you choke their growth and prosperity. You foredoom them to stagnation. You deny them the right to expand their basic industry. You commit a major crime against a great section of our country." O'Sullivan draws upon the imperialistic rhetoric to sway the committee members to vote for the project, using the catch-phrases and arguments that had served the dam boosters well. Without the dam, growth will be "choked" and the states will suffer "stagnation." In contrast, with the building of the dam O'Sullivan associates growth, prosperity, expansion, and progress.

O'Sullivan argued that the region is "an economic unit practically isolated, through long freight hauls, from the rest of the country" and thus needed its own industrial and agricultural infrastructure. "In common, [the farmers] can not possibly make any further progress without reclamation." Reclamation, then, is the motor that can drive industrial and agricultural progress in the West. These states can form an empire that will benefit the nation as a whole, but only given the largess of a majority of members of the U.S. Congress. This largess, of course, is portrayed as a "right" rather than a privilege; not to fund this project would be a "major crime" against the region, and by extension, against the nation. What is clear in this imperialistic rhetoric is O'Sullivan's reliance on a high-modernist conception of the government undertaking the development. It is the duty of the state to build the empire in the Pacific Northwest.

In addition to his testimony at hearings and correspondence with officials, O'Sullivan published many articles in various newspapers around Washington and nationally. O'Sullivan was a focused, perhaps even obsessive promoter of the dam, and his expressive rhetoric was a close match for Woods' capitalized headlines.

In a 1934 article in *The Seattle Star,* O'Sullivan argued against the charges that the Grand Coulee Dam was a relatively new idea. "Coulee Section Always Has Had U.S. Attention" points out that F. H. Newell, Theodore Roosevelt's appointee to the Reclamation Service, had investigated the site more than thirty years prior to the start of construction (O'Sullivan 1934a). This appeal to recent history for present-day legitimacy links up with his general use of imperialistic rhetoric.[4] In the *Star* article, O'Sullivan goes on to tell the story of the dam, with various inaccuracies, up to the point where, in 1934, the low dam construction had begun. In the article, O'Sullivan pushed for immediate construction of the high dam, calling for increased production of power. The low dam, he pointed out, was in part a product of the need for increased production during and after World War I. Although this war was finished, he argued presciently, productive capacity was always necessary in the future:

> The war is over. It may look as if there is not the insistent demand to put these lands to production that there was in 1917, yet those that look into the future can see even a greater demand that these lands be irrigated at the earliest possible time. It must be remembered that it will require some ten years yet before any of the land can be placed to irrigation. We must not look at the necessities of today, but at the necessities of ten years from now.

The future and the past, then, link up to form a linear progression. The great men of history, such as Teddy Roosevelt—and his trusted engineer Newell—had paid attention to Grand Coulee. The project now was coming to fruition, but due to the shortsightedness of present-day proponents of the low dam, we risk hobbling our future productive capacity. O'Sullivan makes a thinly-veiled call for America to be as farsighted as Teddy Roosevelt by building the infrastructure now for a great empire of the future. We can assume that as well as Washingtonians, part of O'Sullivan's intended audience is Teddy Roosevelt's cousin, Franklin D. Roosevelt.

This article was aimed at promoting the project as a whole and the high dam in particular to the citizens of western Washington. Many Seattlites had a dim view of the project, in part because it was a distant and rural part of the state. But a large part of the Seattle opposition came from the superintendent of Seattle City Light, J. D. Ross (this opposition will be discussed in more detail in the next section). Ross fought hard against the Grand Coulee project from the start, perhaps due to the fact that his utility company had plans to dam other sites on the Columbia that would be inundated by the large reservoir created by Grand Coulee. In an article in the *Seattle Times* entitled "City Light Head Sees East Side Domination" (Gilbert 1934) Ross is quoted:

> One official close to Secretary Ickes suggested that Seattle obtain its electrical power from the Coulee dam and forget the Skagit. It should be apparent to every friend of City Light that we must work out our own problems [without the help of the Department of Interior]. The Skagit is the cheapest power in Washington. It is the one heritage we should protect, and by our own efforts offset as far as possible the loss of trade from the East Side due to the development of the Columbia River projects and to the construction of highways leading to Portland.

Through rhetorical devices calling on local pride and competition (the east versus west side of the Cascade Mountains), and "heritage," Ross used newspaper articles and speeches to create a climate of distrust about the Grand Coulee project. He used the individualist orientation to call on Washingtonians to solve their "own problems" through their "own efforts." No governmental interference was necessary in Ross's anti-high-modernist perspective.

O'Sullivan (1934a) saw it as his duty to defend against the attacks, using an expansive imperialistic rhetoric:

> The Columbia Basin Project means much not only to the people of eastern Washington, but to those on the west side. If we put a man on a farm, we must put another man in a small community to serve him, for that is the history of the project, and for each of the men on the farms and for each of the men in the small town, we must have two or more men in the

> cities to provide for them in the manufacturing and mercantile field. So with the great population that is bound to come to the project, the citizens of the coast will benefit two or more fold by virtue of this reclamation. This in turn will build up their shipping, it will create a market for all of their surplus power.

Here, O'Sullivan is being polite. At other times he calls Ross's charges "absurd" or "grossly in error." These comments were always couched in terms of a high-modernist and imperialistic discourse: "As a matter of fact, the Grand Coulee development is absolutely necessary to justify Skagit. The Skagit means nothing but power development. The Columbia Basin development means vast irrigation and new population and that is what we need in the Northwest more than anything else" (O'Sullivan 1934b).

O'Sullivan and Woods were not alone in their promotion of the dam using imperialistic rhetoric. Senator Clarence C. Dill (1934) wrote in *The Seattle Star* of "Lower Cost Of Power To Help Nation," John R. Wheeler (1934) argued that "Projects Along the Columbia To Be World's Greatest," and R. J. Harvison (1936) wrote in the *Seattle Times* of "'Eighth Wonder of the World,' Harnessing Mighty Columbia Steadily Rounding Into Form." The size of Grand Coulee was compared to the Egyptian pyramids and to the Capitol building in Washington, D.C. Images were even printed alongside the article text, with the pyramids or capitol superimposed on an artistic rendering of what the dam would look like (see Figure 5.1). Such photographs show the sheer size of the dam and legitimate the rhetoric used to describe it. But the images do more than this. By superimposing the capitol building against a representation of the dam, the newspapers show the potential power of Grand Coulee Dam.

Figure 5.1 The U.S. Capitol and Grand Coulee Dam, *Wenatchee Daily World* 15 March 1937.

Indeed, reporters eventually reflected on the continual use of this rhetoric: "Any discussion of the Grand Coulee Project must involve a monotonous repetition of superlatives. Few features of it that are not 'bigger,' higher,' 'longer,' greater this or larger that, than anything of the kind known before" (Harvison 1936) (see Figure 5.2).

Figure 5.2 "Eighth Wonder of the World" The Spokane *Spokesman-Review* 26 January 1936. Reprinted with permission of *The Spokesman-Review*.

As it was used to describe and boost Grand Coulee Dam, imperialistic rhetoric combined the Progressive-Era views of linear progress in human history and belief in rational applications of science and technology to solve environmental "problems." Grand Coulee Dam was discursively constructed as the "eighth wonder of the world" and the "biggest thing man has ever built," that is, an unsurpassed technological achievement. The dam represented a domination of nature by humans, but with nature's tacit approval, for nature had shown humans the way to do it. The dam would allow the nation to achieve the grandeur that was predicted by Thomas Jefferson when he made the Louisiana Purchase, to fulfill the hopes of Louis and Clark when they "discovered" the Columbia River and the Pacific Northwest. The dam, in short, represented a final civilizing achievement on the West Coast, consolidating the empire built upon the idea of manifest destiny.

This imperialistic rhetoric was inflected in two ways: by high-modernist impulses toward state-sponsored development and by individualist capitalistic development. Both groups of boosters used the imperialist discourse of empire building and human domination of nature through science and technology. They did not, however, agree on who should do the dominating. The pumping plan advocates realized that the job was too large in scale for either individual capitalists or even the State of Washington to build. They looked to the federal government instead, calling on the U. S. Bureau of Reclamation to support the progress that could achieve an empire in the Pacific Northwest. The gravity plan advocates, on the other hand, worked to keep the federal government out of the Pacific Northwest. They feared the development of cheap public power with wide distribution, and so they pushed for Washingtonians to solve their own problems through their own efforts. This individual orientation eventually lost out to the high-modernist state-sponsored plan.

THE HEBRAIC DISCOURSE: PIONEERS IN A CHOSEN LAND

Imperialism was not the only thing that concerned those who boosted the dam. They also saw the Pacific Northwest in general, and north central Washington in particular, as a special place, made up of special people. Following Weber (1952) and Gorski (2000), I label this sense of being a chosen group an Hebraic discourse. The Hebraic discourse relies upon certain rhetorical strategies: (1) the notion of a chosen people, (2) God smiling on a promised land, and (3) these special inhabitants of the promised land benefiting from a supreme benevolence based on an exclusive *berith* or contract with God (Weber 1952). The local inhabitants of north central Washington, being descended from brave pioneers, struggled to fulfill the richness of the "promised land" and the "last frontier." As it was addressed to different audiences, the Hebraic discourse emphasized the land, the people, or their status. When directed toward other locals, the Hebraic discourse often pointed to the special qualities of the chosen people and the blessed land and resources of the region. When directed at a national audience, this rhetoric often focused more on the status of the region, pointing to its deserving qualities and the entitlement to benefits not yet granted.

Many articles drew upon the rhetoric of a chosen land and a chosen people. These locals were usually labeled as heroic pioneers, the visionaries who proposed the dam and the laborers who built it were lauded in newspapers such as the *Seattle Star,* the *North-Central Labor Advocate,* and the *Seattle Post-Intelligencer,* with articles such as the 1934 "Basin Project Construction Is Victory for Pioneers of State" (Post-Intelligencer 1934).

In a 1934 article in the *Seattle Star,* entitled "Immensity of Grand Coulee Project Hard to Appreciate," the Hebraic discourse is used in describing the pioneers who built the project. "In the heart of this great Grand Coulee Dam development there has been established a number of pioneering towns which are comprised of people known to all staunch Americans as empire builders. They are the types of people who undergo hardships and do so gladly; they are satisfied to await substantial returns on their investments and have the courage of their convictions. They live in a city in its crudest state; they watch its development and eventually are completely lost in the avalanche of progress. . . . Theirs has been an indomitable spirit and it cannot be denied but that the American pioneer is deserving of the respect and admiration of those who make up this great continent" (Star 1934).

The Columbia Basin Irrigation League, the group of boosters working out of the Spokane Chamber of Commerce that backed the "pumping plan," was active in the regional and national discourse for the Columbia Basin project. It published many pamphlets, wrote letters to the Bureau of Reclamation and the Secretary of the Interior, and generally pushed forward their plan for the irrigation project. Their pamphlet entitled "The Columbia Basin Irrigation Project: National economics Justify Columbia Basin Project" published in 1928[5] shows many of the rhetorical devices of imperialistic and Hebraic discourses. The pamphlet "A National and Timely Project," among whose "Factors for Success" include "Reclaiming the Desert," expanding "Civic Life," and claims that "In Potential Wealth Columbia Basin Project Is Richest." As with many publications whose audience is both regional and national, the pamphlet combines the imperialistic and Hebraic discourses.

Rufus Woods' *Wenatchee Daily World* also combined the Hebraic and imperialistic discourses under a masthead calling the region "THE APPLE CAPITAL OF THE WORLD AND IN THE BUCKLE OF THE POWER BELT OF THE GREAT NORTHWEST." As Michael Pollan has argued in *The Botany of Desire,* apples symbolize the fruit of knowledge, and evoke images of the Promised Land: "Re-creating a promised land anywhere in the New World without an apple tree would have been unthinkable" (Pollan 2001: 20). In his (perhaps unintentional) evocation of the Promised Land, Woods linked the two key discourses at Grand Coulee—the imperialistic and the Hebraic. Apples, as representative of Eden, linked Hebraic ideas to the imperialistic rhetoric where the harnessing of Columbia River hydropower would buckle the power belt and complete the inland empire of the Pacific Northwest.

Once the Grand Coulee site was clearly chosen and the pumping plan was a foregone conclusion, the Spokane Chamber of Commerce heartily supported the construction. It published a number of pamphlets that were aimed at increasing tourism in the dam construction area and north central Washington in general. The pamphlet "GRAND COULEE DAM, and its headquarters, SPOKANE" (Figure 5.3) was an attempt to convince tourists

Figure 5.3 "Grand Coulee Dam" Spokane Chamber of Commerce Brochure, 1936.

to spend their vacation in Washington, with the main draw being a view of the construction site at Grand Coulee (Spokane Chamber of Commerce 1936b). The Spokane Chamber of Commerce clearly desired that tourists, "in visiting the *Grand Coulee Dam,* come to *Spokane* first, using this city as base headquarters" (emphasis the original).

The pamphlet calls Spokane a "center of tremendous activity" where you can "see an empire in the making . . . a stupendous engineering enterprise in a geological wonderland." Furthermore, this is the "nation's largest power development, a striking feature of western travel." The pamphlet, in its short history of the dam proposal, claims:

> Men of vision and courage dreamed and worked for half a century for this triumph. The vast tide of the largest river entering the Pacific Ocean from the east and the mighty sweep of hundreds of thousands of acres of fertile but water-hungry land, challenged the imagination of the pioneers. Time and again the enterprise they proposed seemed hopeless, yet the struggle was carried forward indomitable. Notable as the Grand Coulee project will be from an engineering standpoint, it is still more notable as a monument of patriotic zeal and lofty vision.

The authors of the pamphlet made use of both imperialistic and Hebraic discourses, claiming that the dam will contribute to empire building ("will unlock doors to great national wealth") done by pioneers with a "lofty vision" in a "land which offers living at its best."

A 1938 pamphlet was entitled "The Columbia Basin Grand Coulee Project: A Remarkable National Resource that will Contribute Perpetually to the Country's Wealth, Prosperity, and Well-Being" (Spokane Chamber of Commerce 1938). This extensive title says more than enough in itself. The pamphlet is a persuasive piece, making the basic arguments regarding conservation: the waste of water that flows to the ocean, and the need for irrigation and electricity. The pamphlet contains a fascinating argument comparing the need for conservation with the need for national defense: "Dr. Isaiah Bowman, president of Johns Hopkins University, characterizes the waste of water in the United States, and attendant evils, as 'worse than war.' . . . It has long been recognized as an economic waste to allow this dependable water supply to rush unchecked and unused to the ocean while close by lies an area the size of Connecticut awaiting only the touch of moisture to transform it into productive farm land." The pamphlet continues by denying the arguments of many that this new land would compete with farms east of the Rocky Mountains, and points out benefits to employment and feeding hungry people. Furthermore, the dam is "on matching scale with nature" and nature has provided a "ready made canal." The pamphlet shows pictures of sagebrush desert under the caption "potential gardens."

The key to unlocking economic wealth, aside from the dramatic increases in industrial power, would be through tourism. In order to promote this aspect of the dam, the Columbia Basin Commission, a group made up of local and regional boosters, including James O'Sullivan, published a softcover book that describes the benefits of tourism (Columbia Basin Commission 193?).[6] This large, glossy publication described where tourists would go when visiting the dam and what other sites they should visit. The work poured over the details of the tourist sights at the dam, including an electrical palace, the light show, and a 500-foot tall aluminum tower. These tourist attractions overlook the dam site from a location known as Crown Point. This was not simply a tourist brochure though. It was also a persuasive document, with section headings that argued a case for the dam, "how it will pay for itself many times," how the dam had many "benefits to business" in the area, especially "the tourist business." The book described the high level of "food consumption by tourists" and gently reminded area businesspeople of "the tourist reservoir" in the U.S., by which the authors mean a suggested $10 billion spent annually by tourists.

As the very descendents of these brave pioneers whom the boosters lauded, the farmers in the north central Washington area had a more complex relationship to Grand Coulee than one might initially imagine. Farmers' organizations such as the Grange originally opposed the plan to bring irrigation water to the area. The Grange was a group of established and mostly successful farmers. As such, grangers generally felt that there enough land under cultivation already. Following the lead of the Secretary of Agriculture, they argued that to bring more land into production would mean competition with established farmers. Until 1929, the Washington State Grange opposed the project, and, as Elizabeth Sanders (1999) argues, this powerful opposition could have doomed the project. In 1929, however, the Grange reversed its position based on two promises from the Bureau of Reclamation.

The Grange may have opposed increasing the amount of land under production, but it was less opposed to this than it was in favor of cheap public power. In essence, the Grange was seduced by the lure of cheap hydropower from Grand Coulee. Not only could this cheap power light the farmhouses in rural areas, but the Grange had positioned itself in opposition to the power trusts both in Washington and nationally. In exchange for its support for Grand Coulee, the Grange extracted a promise from the Bureau that new lands would not be brought into production until they were needed, and this timeframe was estimated to be perhaps 30 years (Pitzer 1994). Soon after this state-level proclamation, local Granges from all over Oregon and Washington began to send letters of support to the Department of Interior and to Congress.

The support of the Grange was one of the turning points in the project. With farmers fully behind the pumping plan, gravity plan boosters began to see support for their project falter. Not one to let the support of the farmers go unnoticed, Roy R. Gill, a hardware merchant from Spokane, and the executive chairman of the Columbia Basin Irrigation League (a Spokane Chamber of Commerce group) helped spread the word that farmers were behind him. At the House of Representatives Committee on Irrigation and Reclamation Hearings in June, 1932, Gill read a survey of all the landowners within the area of the project. One question was, "'Are you in favor of irrigation of the Columbia Basin project?' Of the replies received, 91 percent were 'yes'; 4 percent were 'no'; unanswered; 5 per cent. That shows that the farmers of that country are pressing for this project" (House of Representatives 1932: 222). Gill used this survey of landowners to support the pumping plan once it appeared to be a foregone conclusion.

The role of the local Indian tribes in the discourse around Grand Coulee is similarly complex. The native peoples living in the Big Bend region were a diverse collection of tribes. The tribes most closely affected by the dam are grouped together onto one reservation, and make up one governance unit, the Confederated Tribes of the Colville Reservation. The Colville Reservation, established by President Grant in 1872, contains about 8,000 members in 12 tribes, four of which were directly affected by the dam: the San Poil, Colville, Nespelam, and the Lakes. The other tribes are also from the dam's immediate region, including the Okanogan, Methow, Columbia, Wenatchee, Chelan, Entiat, and the Palus. Chief Joseph's band of the Nez Perce also moved to the reservation from its native lands in the Montana area. At the very upper reaches of the reservoir, the Spokane Indian Reservation was also affected.

Indians objectively had the most to lose. Each of the two reservations lost about 3,000 acres of land to the reservoir. Two towns on the Colville Reservation had to be relocated to avoid inundation, and the salmon runs that the Colville relied upon for their livelihood were decimated. These groups, however, also had a great deal to gain. The government promised access to hydropower, some amount of proceeds from the sale of hydropower, and water for both drinking and irrigation.

Though many benefits of the dam were promised, only the Spokane Reservation received any remuneration for their lost land. They were not, however, involved with the dam to the extent that the tribes of the Colville Reservation were. The Confederated Tribes of the Colville Reservation never received any irrigation water, hydropower, or profit from the sale of the hydropower. They did not receive any recompense for their lost land, the relocation of the towns, or the loss of the salmon (Joseph 2000).

The main interest that the Colville tribes had in the dam was the rights to the river water, and therefore its potential hydropower production. According to the Federal Power Commission, the taking of those rights by the federal government had to be compensated. Harold Ickes directed the Bureau of Indian Affairs, in a letter of 5 December 1933 to "make a recommendation of the amount of revenue that should accrue annually to the Indians by reason of their interests which are involved in the development." This was calculated to be a $63,000 one-time payment. According to the Colville tribe, further payments were promised, but "with the completion of the project and the coming of World War II, this was never done" (Confederated Tribes of the Colville Reservation 1994: 4).

This prompted a lawsuit in 1951 that was not settled until 1994 (Confederated Tribes of the Colville Reservation 1994). The suit exposed the broken promises, and asked compensation for the losses. In the discussions of settling the lawsuit out of court, Eddie Palmanteer testified before the House Committee on Natural Resources in 1994. The dam, Palmanteer said, had been a boon to the State of Washington, but a disaster to the Colville Tribes:

> It irrigated hundreds of thousands of acres, but none on the Colville Reservation. Its construction denied the Colville Tribes the ability to use that site or any other site for its own power development. The dam destroyed forever the mighty salmon runs that used to migrate to Kettle Falls on the upper Columbia River and to the tributaries of the Reservation.

Palmanteer argued that the dam had changed the lives of the Colville tribes forever, and they were entitled to repayment for their losses. They settled the case based on a complex calculation that measures the value of the lost salmon runs based on an estimated volume equivalent to tins of tuna (Joseph 2000).

During the debates over the construction of Grand Coulee and through construction, predictably for the time period, the perspective of the Colville tribes was rarely expressed in the press or in congressional hearings, indicating the neglect of the Anglo government and media at the time. The actions of Indian leaders, however, do indicate a perspective, for tribal elders participated in the ceremonies at both the ground-breaking (see Figure 5.4) and an Indian official even blessed the dam at its dedication.

Figure 5.4 Grand Coulee ground breaking *Wenatchee Daily World*, 16 July 1933.

This tacit approval on the part of some of the leadership did not, however, translate into widespread acceptance on the part of the larger population on the reservation and in the region. As with other native groups at the time (White 1983), because the Colville Reservation is such a diverse population, with 12 distinct tribes with different systems of organization, there were problems with any attempts to control the reservation and commit to help with projects on a legitimate democratic basis. Though there is little recorded history of the perspectives of Indians of the Colville Reservation, Indians who lived through the debates over Grand Coulee point out several factors that complicate an easy interpretation of Indians being supportive of the dam because some leaders cooperated with the U.S. government.

First, during this time there was very little participatory discussion between the tribal government and the reservation population about whether or not the tribes would support the dam. The decision was made by a small group of elders who did not necessarily reflect or consider the wishes of the wider Indian population of the reservation. Hence there was very little contestation of the reservation leader's choices. Second, just as with the farmers, the lure of good land, cheap water and hydropower, and Depression-era jobs had a role in muting any opposition. As expressed by Paul James (2000), "Some did OK and some didn't. . . . They were promised a lot of things that never did come to reality." Clearly, the Hebraic discourses about north central Washington being a promised land was not intended to include the native inhabitants.

Hebraic discourse about Grand Coulee Dam glorified the north central Washington region and its inhabitants. Calling the land "rich," "vast," and "of incredible scenic beauty," pamphlets and newspaper reports constructed the region as a present-day land of milk and honey that needed only a drop of water to burst into bloom. This rhetoric was often combined with imperialistic discourses about building the region into an empire. The Grand Coulee Dam would fulfill the promise of the Promised Land: an empire in the Pacific Northwest. In the eyes of high modernists such as Rufus Woods and James O'Sullivan, it was the federal government who should build Grand Coulee Dam so as to water the land and fulfill the promise of the region. To the individualist capitalists, such as the Spokane Chamber of Commerce, the intrusion of the government was unwanted; the land could be reclaimed solely through the efforts of the courageous men of vision who lived there.

NATIONAL DISCOURSE ABOUT GRAND COULEE: 'THE EIGHTH WORLD WONDER'

National discourse around Grand Coulee Dam made use of imperialistic discourses to highlight the spectacle of the dam (it would be the "eighth wonder of the world") and the economic improvements that it would bring to the region and the nation as a whole. Predictably, national press had little interest in glorifying the local inhabitants by using Hebraic discourses. Instead, the national discourse about Grand Coulee focused on the ways that the dam might, along with improving the local and national economy, improve the local inhabitants, a discourse that is best described as anti-Hebraic. The anti-Hebraic and high modernist imperialistic discourses were used by the national press to great effect: the dam would be the capstone in the process of making the United States an economic and industrial empire. In the process the citizens of the frontier would be civilized and the whole nation would benefit (similar arguments about reclamation projects can be found in Pisani 1983; Scott Unpublished Manuscript: 20).

Rather than a local booster, it was a writer for a national periodical that coined the phrase that lodged in the national memory. Robert Ormond Case, writing for the *Saturday Evening Post,* entitled his 1935 article "The

Eighth World Wonder," and went on to describe the dam in enthusiastic terms (Case 1935):

> Today, as this is written, the fifty-century record is going by the boards. The ground-work of a structure that will transcend the mass of the Great Pyramid is being laid. The tomb of Cheops served no useful purpose, in the utilitarian sense of the word; when it was completed, the sand above which it towered remained as barren as before. The modern structure at least has the benefit of an economic dream behind it: It is designed to water an area greater than the delta of the Nile. . . . This modern eighth wonder of the world—the structure that will overshadow for the first time in 5000 years the stupendous folly of the Pharaohs—is the dam of the Grand Coulee power and reclamation project on the Columbia River.

Case used the same superlatives, with the same reflective commentary, that local Washington State newspapers used: "Superlatives founded on fact come readily to the tongue when speaking of Grand Coulee. Figures are at hand that command attention even in a ten-digit era." The article described the project in great detail, giving both a "bird's eye view" and a "worm's eye view" in order to edify the national reading public. Case's rhetoric made use of the imperialistic devices of the local boosters, yet tempered it with some discussion of the contentious aspects of the dam. Though he compared the dam favorably with the great Egyptian pyramids, Case asked the question that seemed to be on everyone's lips except James O'Sullivan—*who will use the power?* "As to the feasibility of the project . . . who will pay, and how . . . or who will use the 2,600,000-odd horsepower of available energy." Case was referring to the fact that when Grand Coulee was built there was no demand for extra power. In fact, the recent approval of Bonneville Dam was justified by the argument that it would provide *surplus* power for the region for many years to come. Building Grand Coulee would provide so much cheap power that local utilities would have been undercut by the government—exactly the fear of the local elites who backed the gravity plan.

Although Case was critical about the surplus power production, he lauded the promoters of the dam:

> It was the spectacle of this vast natural reservoir, adjacent to the river and situated above more than 1,000,000 acres of fertile land awaiting water that caused the vision of reclamation to germinate, bud and grow. One is at once struck by the magnitude of the dream, the sheer size of the units that comprise the composite picture. The proponents of the project—and they are many—obviously recognize no limit to human endeavor, admit no farther horizon to human accomplishment.

Though Case was ambiguous about the extent of his identification with the proponents of the dam, his description of them took the high modernist imperialistic rhetoric—where progress is without limit—to an extreme, in some ways outdoing Rufus Woods' own exaggeration.

In the national media, however large the dam was and however well it compared to the great pyramids, a question kept coming up: "Is it worth it?" Over and over, critics of the New Deal programs blasted the President and his staff for several related issues. First was the surplus power question. Second was the method of appropriation; because FDR had initiated construction without congressional approval by using PWA funds, critics charged him with being anti-democratic. Third, many critics pointed to Reclamation's track record: a bevy of bankrupt farmers, too many expensive dams that appeared to fund only land speculation schemes, and relatively few successful projects. Given these three problems of dubious benefit, the overall question was whether the dam was worth the massive expenditure? Articles that raised these questions were printed in *The Nation, Collier's,* and *Harper's.*

Owen P. White, writing for *Collier's* in 1934, noticed that, with all of the federal money being poured into reclamation, it was primarily the land speculator who profits, "not the farmer. Past records show that he usually goes broke on irrigated land. There will be a rise in land prices, of course, but the speculator will get that. Even the projects that Congress has condemned are being pushed—with PWA funds" (White 1934: 10). This article ran above two cartoons, one of a farmer beginning to sit down in a chair covered with cactus, labeled "reclaimed land crops." The second and larger cartoon showed the U.S. Treasury building reconfigured into a hand pump with the U.S. tax payers dangling from the open valve. The water coming from the treasury flowed into a stream in the shape of a dollar sign, with an overweight man in coat and top hat, labeled as a "land speculator," pulling fat fish out of the stream. A cartoon on a later page shows a dilapidated car labeled "Congress," with a driver who, looking over his shoulder, sees a grinning young couple, labeled as the "Reclamation Bureau" in a shiny car labeled "PWA Funds" driving past a sign that says "to the desert."

Writing in another issue of *Collier's* about the surplus of power, Walter Davenport suggested that "building for the future" was a ridiculous notion given the massive level of production: "Who is going to use the power generated at Grand Coulee? Or at Seminoe, Casper-Alcova, Wheeler, and Pickwick Landing? We're building huge power factories, but no transmission lines. Most of the plants are in regions thinly populated. Markets for government power will have to be created by shiftings of people and industry" (Davenport 1935: 10). The centerpiece of the first page of the article is an unflattering photograph of the construction site: barren cliff walls overlook gravel roads on the muddy floor of the cofferdam. This photograph appears to belie the boosters' claim that such a site could be the eighth wonder of the world.

Davenport's article, while it does not directly announce its bias, goes on to describe the project in fairly negative terms. Sub-sections are entitled "Real Estate Ballyhoo" and "Doubtful Prosperity," and Davenport suggests that the generators inside the dam will be "singing their Lorelei song to cynical industry:"

> These shining monsters will be able to grind out 1,900,000 kilowatts of electric power. Below, from the face of simple dikes that will impound the boiling flood, miles of canals will meander southward, like watery tentacles exploring and groping, aimed at the redemption of a million and a quarter acres of dead land bitter with alkali.

The imagery of the octopus, used against Boulder Dam earlier, evokes the suspect powers of government. James Rorty, of *The Nation,* also wrote about land speculation around the town called Grand Coulee, next to the dam construction site. Rorty (1935) ironically asked:

> Have you ever seen the American Dream walking? Well, I have. I saw it walking up the side of the Columbia River canyon, scribbling its puny etchings of squalor and cupidity against an austere backdrop of leaning cliffs and sudden chasms, and crooning the old American theme songs of Get Rich Quick and Something For Nothing. The dream is a town. It calls itself Grand Coulee; it is built of faith, hope, barn siding, and paperboard; when I was there it was inhabited by about 1,500 people. It had twenty eating places, as many saloons, at least a half-dozen wide-open brothels, five grocery stores, two jewelry stores, a furniture store, two drugstores, two ladies' wear shoppes, three beauty shoppes, a proportionate quota of painless dentists and radio repair shops, and six real-estate agents.

Rorty goes on to blast the President's ideas about relocating poor folks who farm sub-marginal land into this new region. Given the amount of land speculation occurring, government condemnation and distribution of the land would be impossible, for no one knows at what "present value" the land should be condemned. Rorty concludes, with great facetiousness, that perhaps creating a "self-sustaining agricultural economy" wasn't really the idea after all:

> But maybe this wasn't the idea. Maybe the idea was to create a new frontier, an artificially made, publicly financed block of exploitable resources so that the American Dream of Get Rich Quick and Something

> For Nothing might be dreamed all over again. In that case it is proper to remark that the American Dream is obsolete; that there is nothing in it any more except headaches, ultimately, for everybody.

Rorty's critique both blasts the undemocratic method of appropriation for the dam and also pokes fun at Roosevelt's New Deal promises and their potential to be undone by speculators and dreamers. His alone was a total critique of the project, from conception, to appropriation measures, to the building of the dam and a disdain for the dam as a monument. His arguments, though valid, represented a lone voice on the national level. Most commentators, though perhaps critical of one aspect or another, were generally laudatory of the dam.

Richard Neuberger, a writer in residence at the Portland *Oregonian* while he covered the building of Grand Coulee Dam and later a U.S. Representative from Oregon, was a persistent but fair critic. During the early years of dam construction Neuberger published several essays, the first three of which were reprinted in *Our Promised Land* (Neuberger 1938): "Last Frontier," "Promised Land," and "The Biggest Thing on Earth." These articles are clearly a series, with the first two setting up the punch-line in the third. "Last Frontier" described in great detail the potential productive power of the Pacific Northwest: "the Columbia Basin can generate more hydroelectricity, grow more wheat, and cut more timber than England, France, and Germany combined. . . . [a] treasure-trove of undeveloped natural resources of an area more than twice as large as Spain . . . The Pacific Northwest is still largely a frontier."

In "Promised Land," Neuberger described the exodusters who traveled to the Pacific Northwest in search of work, farms, and land: "One need only park alongside a through highway on the coast to see for oneself the caravans of these modern pioneers. They roll westward almost like a parade. . . . The automobiles of the wandering farmers from the Dust Bowl dwindled in the distance. I watched them becoming specks against the grim lava cliffs towering above the highway. Probably the wagons of the forty-niners had looked exactly like that against the same cliffs." Neuberger attacked the President's comments about the Pacific Northwest as a land of opportunity for the Dust Bowl refugees: "Opportunity? For some, but not for everyone . . . the plight of these pilgrims takes many forms." Neuberger describes the forms of this plight in horrendous detail: the squalor of the work camps, the lines for the relief rolls, the bankrupt schools, and the farm-girls-turned-prostitutes.

This ironic description of the Last Frontier and the Promised Land as one inhabited by destitute refugees begs the question that Neuberger asks on the last page: These Americans "must be returned to productive farming if they are to be self-supporting again. Most of them will be taken care of. But what of the pilgrims yet to come? How the government proposes to settle these wandering farmers is the story of the boldest and most daring construction enterprise of all time. It is the story of Grand Coulee." Neuberger, in other words, has set the reader up to see the solution to the horror of the Great Depression in Grand Coulee Dam, "The Biggest Thing on Earth."

This third essay in the series, printed in *Harper's* in February, 1937, praised the dam for solving many of the problems of the Dust Bowl. It would provide new lands for the pilgrims, hydroelectric power for industry, and—perhaps most importantly—much-needed jobs for laborers. "The Biggest Thing on Earth" was quite laudatory of the project overall, but also asked some pointed question of the President's motivations for the project, and pointed out some practical difficulties with these stated goals:

> It is the irrigation phase of Grand Coulee which raises social and economic questions transcending in significance those identified with any other Public Works enterprise of the Roosevelt Administration, except possibly the TVA. The completion of the dam will add more than a million and a half acres to the productive capacity of the nation. This land can be used and developed only if the Federal Government enters into an extensive program of assisting people in the slums and tenements of the East and in the Dust Bowl of the Middle West to settle and cultivate a great chunk of fertile soil almost a continent removed from their homes.

The discussion of Grand Coulee in the national press was, then, decidedly two-sided. While some authors praised the dam as a monument to American ingenuity and know-how, others condemned it as a boondoggle and a ballyhoo: the worst embodiment of the American Dream. When it was employed, the imperialistic discursive strategy was quite similar to its usage at a local level. Boosters on the national level used the rhetoric of a Pacific Northwest empire; Grand Coulee as the fulfillment of economic progress; the dam was also portrayed as the monumental achievement of scientific and technological advancements.

The Pioneer rhetoric of the Hebraic discourse, however, was not used often at the national level. Instead, local inhabitants of the region were often portrayed as simpleton farmers or greedy land speculators; this is an anti-Hebraic discourse. Interestingly, during the period of the New Deal, ideas of social engineering were popular (Pisani 1983), so this anti-Hebraic discourse had a paradoxical power. By portraying the local residents as either needy exodusters or incompetent dirt farmers, the national press held open the door for FDR's social engineering programs that would turn around the backward rural folks.

Whether Imperialistic or anti-Hebraic, national-level discourses about Grand Coulee Dam helped to keep the project on the agenda of the government at times when it may otherwise have been an uphill battle for the local boosters. The national press succeeded in sensationalizing the dam by showing its economic might, its massive size, and the tremendous economic and social consequences of its construction. By keeping the dam in the national attention with generally laudatory terms, the national press contributed to the juggernaut of Grand Coulee Dam. Without a sustained critique from the national press, Grand Coulee Dam had no enemies. Though reporters sometimes questioned the way that the project was funded, or the manner of distributing surplus power, they did not push these issues into a full critique. Instead, they pointed out the problems, but many then immediately asked how any good Americans could possibly turn down the project that was to be the 'biggest man-made thing in the world'? Americans could not, especially since such an achievement would help civilize the backward sections of the country. Only Rorty stands out as unwilling to portray the dam as, in the final analysis, a good thing. But without a chorus of voices, Rorty could not sustain a full critique alone. Because the national press did not find damning fault with the Grand Coulee, the national press functioned as another supportive factor in constructing the dam.

GOVERNMENTAL DISCOURSE I: CONGRESS, THE CORPS, AND THE BUREAU

The representation of locals as hicks (or worse) carried through to discourse within the government as well. While FDR infamously referred to the social engineering aspects of the TVA (Roosevelt wanted the TVA to bring the region "into the twentieth century," to "make a different type of citizen . . . not only the mountaineers—we all know about them—but it applies to the people around muscle shoals" [quoted in Scott unpublished manuscript: 20]), other agencies within the government paid little attention to the population of eastern Washington. The Bureau of Reclamation (within the Department of Interior), Army Corps of Engineers (within the Department of Defense), and Congress all debated the dam during its various stages, received and responded to letters from concerned citizens, and engaged in debates both with each other and within the confines of the individual governmental bodies. This governmental discourse, however, focuses almost entirely on imperialistic rhetorical constructions of the Grand Coulee project. The high modernist imperialistic rhetoric of the government relies on the same discourse used by local boosters: constructions of the dam painted it as the crucial piece of an industrial empire that will contribute to important national concerns, such as national defense and economic growth.

The imperialistic discourse at the governmental level abounds, especially in committee hearings and in the text of legislation. The House and Senate each had a Committee on Irrigation and Reclamation that heard testimony on proposed irrigation projects. In these hearings one can see the imperialistic discourse of empire building.

The 69th Congress held hearings on 2 February 1926, on the question of allocations for Columbia Basin research. The president of the Oregon Irrigation Congress, James M. Kyle, used typical imperialistic rhetoric: "This new empire, carved from the sagebrush, will create an extensive buyers' market for Eastern and Middle Western manufactured products." Kyle makes clear reference to empire building and implies a courageous "carving out" of this empire from useless terrain made only of sagebrush (Senate 1926: 21).

The House of Representative's committee hearings of the 72nd Congress debated the Columbia Basin Project Act in May and June of 1932. This act would eventually fund the entire Grand Coulee project. Commissioner of Reclamation Elwood Mead wrote a letter to the committee, saying (House of Representatives 1932): "As a conservation measure this ranks among the greatest yet attempted in this country. The water of the Columbia River is the most valuable resource of the Northwest. It has the immeasurable value of being permanent, of being made the basis of a great industrial and agricultural development." Again, the imperialistic rhetoric invoked the water project as a foundation for future development that can only be described in superlative terms: "great," "stupendous," and "outstanding."

During the same hearing Representative Scott Leavitt, Republican of Montana suggested (House of Representatives 1932):

> The question of national defense in connections with the Panama Canal was raised; and it has always struck me, in connection with this project, that it has, from a national defense standpoint, a tremendous value. The orderly development of our population and resources on the west coast is of extreme importance; and the way to develop is through the development of those irrigated sections that become the center of all industry and live stock raising and so on. We have here, of course, a combination of power development and the development of a great area for tremendously aiding the growth of population on the coast. We will never be fully developed, from a national defense standpoint, until the West Coast is as fully developed as the East Coast.

Leavitt's comments encapsulated the central points of imperialistic discourse. The development of the Columbia River was to be made for the economic progress of the region and the nation, for national defense, for the orderly settlement of the region, and the increase of the local population. The strategic concerns of state building are laid bare in Leavitt's speech; economic development, political and military strengthening, and settling of the lands with the proper kinds of people are the goals of the Columbia Basin project.

The text of legislation also contained traces of imperialistic discourse within the straitjacket of legalistic language. One bill, HR 1434, which linked the power output of Grand Coulee and Bonneville Dams, passed 6 January 1941:

> Be it enacted by the Senate and House of Representatives of the United States of America in Congress assembled, That the Pacific Northwest is a vast storehouse of natural resources which can provide products useful for the general welfare of the people in times of peace and critical and strategic materials for the national defense in times of war. Through the construction of extensive facilities for the control and conservation of the water of the Columbia River, the Federal government has made possible the greater utilization of these rich resources.

The region was seen as a vast storehouse, crucial to the needs of the nation. Natural resources were useful for the general welfare of the people in peacetime, but were also to be developed for national defense.

As opposed to committee hearings, which were more insulated from public discourse, much of the correspondence within the Department of the Interior centered on controversies brought up in the national media. Rorty's article in *The Nation* was a particular thorn in the side of many administrators. James O'Sullivan, C.C. Dill, and others took it upon themselves to write to many officials, defending the dam. Much of this official correspondence carried on the themes of the imperialistic discourse. Harry Slattery, the Secretary of the Federal Emergency Relief Administration at the time, responded to C. C. Dill. In his letter, Slattery called Dill's work on the project "disinterested but effective," saying that he shared Dill's resentment over what they both perceived as misrepresentations in Rorty's article: "The water of the Columbia River is one of the great assets of this nation, capable of creating wealth and ministering to human comfort to a greater degree than any other undeveloped resource."[7] Slattery's comments show his view of the region as an integral part of the nation, with the Columbia providing resources to the rest of the country, and the government providing resources to develop those resources.

The Bureau of Reclamation and the Army Corps of Engineers also contributed to the imperialistic discourse on Grand Coulee through pamphlets, reports, and engineering studies. In a strange competition between government agencies, both the Bureau and the Corps studied the Columbia River and proposed a dam at Grand Coulee, as discussed earlier. The documents of both of these agencies, but especially those of the Bureau of Reclamation, show the imperialistic high modernist discourse.

The Corps made several studies of the Columbia River, and in 1932 had begun construction on the 200 foot-high Bonneville Dam near Portland, Oregon. On the upper reaches of the Columbia the Corps recommended building several dams to control flooding and to improve navigation. In 1918 the Corps had no interest in the project due to technical barriers (the 4,000 foot-wide and 700 foot-deep canyon), but by 1932 it was eager to gain new contracts. Major Butler, author of the "308 Report" that recommended multiple dams on the Columbia, testified before the House Committee on Irrigation and Reclamation on 1 June 1932. He claimed that a dam at Grand Coulee was both possible and desirable (House of Representatives 1932: 17):

> The pumping plan of placing water on the Columbia Basin project is shown as being altogether feasible both from an economic and an engineering viewpoint. This plan provides for the construction of a high dam at the Grand Coulee site on Columbia River of such height as to back the water up to the Canadian border and thus utilize all the available head.

Butler was arguing for a dam that would please the boosters such as James O'Sullivan and Rufus Woods—a high dam that would create an expansive reservoir reaching to the Canadian border. Butler continued (House of Representatives 1932: 19):

> The complete, or even partial development of this stream is a stupendous undertaking, involving outstanding engineering and economic problems, although they are by no means infeasible or insurmountable.

Butler's comments showed elements of the imperialistic rhetoric; the dam was stupendous, the challenge was outstanding. The Corps played a roll in helping gain Congressional support for the dam, but the construction responsibilities were eventually granted to the Bureau of Reclamation, within the Department of Interior, where Harold Ickes could directly oversee it as one of the New Deal public works projects.

The Bureau of Reclamation also had an ambivalent relationship to the Grand Coulee Dam in its early history. While the Corps rejected the dam due to infeasibility in the early years, the Bureau did not pursue it largely due to the influence of Elwood Mead and Arthur Powell Davis, who were attempting to slim-down the Bureau in the early 1920s.[8] With the election of Roosevelt and the beginnings of the New Deal, the Bureau came to recognize Grand Coulee as a project that could help the local region, traveling workers, and the economy as a whole. Their rhetoric in arguing for the dam showed such imperialist and high modernist ideology of linear progress, scientific management, and the benevolence of the state in building-up the region for the multiple purposes of social control, the stimulation of industry and agriculture, and the extension of civilizing benefits.

The Bureau produced a pamphlet entitled "Putting the Columbia River to Work" in 1938 that made claims about these many benefits that the Grand Coulee Dam and the Columbia Basin Project would bring. The dam, the pamphlet argued, would provide water to the rich agricultural lands of the area, fulfilling the dreams of the earliest settlers. Following a line of thought that FDR had spoken on, the pamphlet declared that the opening of new rich farmlands in the Pacific Northwest would help solve problems of the Dust Bowl. Furthermore, these benefits would spread to other areas of the nation as the farmers bought materials from Midwestern and Eastern U.S. industries and distribute their produce back to the Midwest, East and South (Bureau of Reclamation 1938: 1):

> The use of water from the Columbia River for the irrigating of enormous tracts of rich land in Central Washington has been dreamed of since the day of the earliest settlers. Modern equipment and methods, electrical power, and Federal loans make the Columbia Project feasible. The enforced abandonment of eroded, sub marginal, and 'dust bowl' lands, the natural increase in population, and the desire for improvement in standards of living make it necessary. . . . The Western States can never be agriculturally self-supporting. They are now and always will be heavy purchasers of eastern agricultural products. Their own agricultural products supplement rather than compete with eastern crops in producing a balanced national diet.

The authors of the pamphlet described the benefits that would accrue to the region and the nation, all goals of state building: the improvement of farm lands, the increase of the population, the increasing standard of living.

The Bureau of Reclamation also trumpeted its achievements in press releases at regular intervals. The press releases[9] showed similar enthusiasm for the development of the Columbia River. On 17 December 1936, when the first construction was undertaken, the Bureau proclaimed:

> Government engineers have pushed the Columbia River, the second largest in America, out of its bed. It is the first time in the history of the

> world man has moved this giant stream which was shoved across a bit of landscape in eastern Washington to make way for builders of Grand Coulee Dam, the biggest dam undertaken to date. The last time the Columbia River was pushed around was about 10,000 years ago in the ice age when glaciers descended over the continent and diverted the Columbia at about the same point. Taking advantage of this start of the work nature made a hundred centuries past, man is now harnessing the torrent to make it serve for power generation and irrigation purposes.

The Bureau was supremely confident of their ability to turn nature to society's ends of hydropower production, farming, and the stimulation of commerce and population growth. Comparing the work of governmental engineers to the great glaciers that helped carve the coulees in Big Bend country gave a natural, inevitable quality to the work being done on the Columbia River.

On 22 March 1941 the Bureau reported the first electric power generated by the dam:

> Initial generation of electric energy today by Grand Coulee Dam on the Columbia River was hailed as an aid to national defense by President Franklin D. Roosevelt. In a letter to Frank A. Banks, Supervising Engineer of the giant project, President Roosevelt declared that putting the dam to work two years ahead of schedule "is a fine job well done." The project, the President said, "served to provide much useful employment at a time eight years ago when it was important that we find at once a means of avoiding complete economic stagnation, and it will serve now to provide the power to make aluminum for airplanes and otherwise to speed our protective arms." Grand Coulee Dam, the greatest structure ever erected by man, will be the largest source of hydroelectric power in the world when all generators have been installed.

Using FDR's words, the Bureau elevated its work to the level of national defense. Not only did the dam help achieve the goals of state building, such as stimulating economic progress and providing resources for the population of the region, these goals were expanded to include the war effort. The hyperbolic phrasing of the press release showed the pride with which the Bureau dammed the Columbia: Grand Coulee is the "greatest" and the "largest." The elements of the imperialistic rhetoric in these press releases were centered on the role that the government played in achieving the victory of nature, and erecting a monument to progress, all in the name of producing an empire in the Pacific Northwest.

Such an empire was seen as a direct lineage from the great civilizations of Egypt, China, India, and Italy. In a pamphlet produced to tell the story of the 100,000-acre irrigation project that went along with the dam, the Bureau of Reclamation again showed such high modernist rhetoric (Bureau of Reclamation 1964: ix):

> Years later [after the Egyptians and Indians] this same continuing quest for water to make the land productive was the driving force behind the building of Grand Coulee Dam. The electric power potential of the proposed dam was secondary. Primarily it was the belief that water from the Columbia River would turn the desert lands of the Columbia Basin into a prosperous and productive region that was responsible for the building of Grand Coulee Dam and the huge Columbia Basin Irrigation Project.

The goals of creating a "prosperous and productive region" out of a sagebrush desert were seen as a linear development from the great civilizations of the past to the pioneers of recent history.

In the pamphlet, the Bureau goes to great lengths to show the national (and even international gains) from Grand Coulee Dam (Bureau of Reclamation 1964: iv):

> The many benefits from the Columbia Basin Project know no boundaries. The San Francisco businessman who lunches on split-pea soup, the Chicago housewife who shops for baking potatoes, the New York hostess who serves after dinner mints—all may be affected in some measure by the produce of the Columbia River Project. The worker in the bauxite mines of Madagascar, the mechanic in a farm machinery factory in Illinois, the freight handler in a Michigan warehouse, the steel mill worker in Ohio, all find a reflection in their weekly paycheck of the expanded economy and purchasing power resulting form the Columbia River Project.

The Bureau of Reclamation and Congress described the ways that the Grand Coulee project would help build an empire in the Pacific Northwest by harnessing the power of the Columbia River; an empire that would benefit the entire nation, and even other parts of the world. This would help carve a great industrial power out of what is now simply sagebrush, they argued, and would contribute to national defense in the same way as the Panama Canal. Governmental discourse at this level virtually ignored the population of the region, and engaged solely in high modernist imperialistic discourse.

GOVERNMENTAL DISCOURSE II: THE PRESIDENT AND THE PROTESTOR

Franklin D. Roosevelt, on the other hand, did not ignore the local population—he visited the Pacific Northwest many times in the years leading up to his presidency. His concern for the locals is recorded in the speeches he made, both during the 1932 campaign and as President. As with the other national-level discussions, FDR also relied on the discourse of imperialism, connecting the needs of the West with the needs of the nation as a whole. One of FDR's political strengths was his ability to include many diverse groups in his national agenda, and this is evident in his speeches and letters from as early as 1920 through his visit to Grand Coulee in 1934.

In a speech at Billings, Montana, 17 August 1920, FDR said:

> The West is progressive. If it had not been progressive there would be no West as we know it today. Every time I come back to these Mountain States I see changes take place in two or three years which would require a generation in the older parts of the Nation—and yet you in the West have but scratched the surface. You want to know, quite properly, what your government is going to do to aid in the continuation of that progress. You want to know the theory of future development, and the purpose behind those in charge at Washington. . . . The use of public lands, the undertaking of new power plants, the bringing in of new acreage of agriculture, all of these are of national importance.[10]

Roosevelt continued in this vein, two days later at the State Armory in Spokane, Washington:

> The territory embraced in the so-called Columbia Basin project is, for example, one of first national importance. Washington is not the only State concerned, either in its building or its future benefits. It belongs to the whole Nation, and the Federal Government must cooperate with the Northwest in plans of such splendid magnitude as that.[11]

FDR was simultaneously relying on imperialistic rhetoric about progress and the American empire, while also drawing on the myth of his cousin Theodore Roosevelt and the Progressive politicians of the era that was just coming to a close. In this speech, and through out this period of his career, FDR outlined his new style of progressivism, one which combined the economic might of the federal government with the ideas of linear progress, scientific rationality, and technical solutions to social problems.

FDR echoed these thoughts in a speech at Grand Coulee in 1934 (as recorded in Walla-Walla Union-Bulletin 1934):

> When you cross the Mountain States and that portion of the Coast states that lie well back from the ocean you are impressed by those great stretches of physical territory, territory now practically untouched but destined some day to contain the homes of thousands of citizens, a territory to be

> developed by the nation and for the nation. As we were coming down the river today, I could not help but think, as everyone does, of all that water running down to the sea.

To FDR, letting the waters of the Columbia run to the sea was a waste. Instead, he envisioned widespread development of the area through reclamation: progress that would benefit the entire nation, not just a remote corner. He told the crowd:

> A great many years ago, 75 or 80, a great editor in New York said: 'Go West, young man, go West.' Horace Greeley is supposed to be out of date today, but there is a great opportunity for the people in the East, people in the South and in some of the over-crowded parts of the Middle West–some people from sub-marginal lands . . . shall have the opportunity of still going West.

This continual referencing of the power of the West to fix the social ills of the nation is a hallmark of FDR's imperialistic rhetoric. The development of the West, specifically the hydropower and reclamation aspects, were essential to the economic health of the nation. They thus take on an important role in FDR's New Deal.

Harold Ickes, FDR's Secretary of the Interior, consistently pushed the New Deal program both as Secretary of Interior and his contemporaneous position as Administrator of Public Works. Ickes argued for expansion of public works projects for multiple purposes. In his speeches, letters, and published articles, Ickes presented public works such as Grand Coulee as crucial for America's success.

In an article printed in the New York *Times* on 12 May 1935, entitled "Public Works for Social Gain," Ickes defends public works projects using the imperialistic discourse:

> The U.S. has entered upon the largest works program carried in a single appropriation ever undertaken by this or any other country in a time of peace . . . it reaches proportions which must make such enterprises as the Pyramids and the construction of the Great Wall of China seem almost petty. . . . We have scarcely begun to take full advantage of the infinite possibilities of our country. We have scarcely begun the work of protecting it from the physical dangers, some caused by our own folly, some created by our failure to control the forces of nature that threaten it. . . . We have passed the stage when we can safely leave them [natural resources] to nature and to the give and take of economic forces. We must interfere with nature in order to give her a chance to serve us. We must limit reckless exploitation in order to preserve the fundamentals of American individualism.

In this striking passage, Ickes used the imperialistic discourse to compare Grand Coulee to the monuments of great ancient civilizations. He argued that some form of governmental protection was necessary, and that nature must be altered if humans are to benefit from natural resources and avoid natural harms.

Warning against waste, Ickes wanted to make wise use of natural resources. His method of developing resources while being mindful of conservation would bring the greatest good to the greatest number: "Within the boundaries of the United States and its possessions, there are resources of almost incalculable value. In the past we have been prodigal with our natural wealth . . . But while we have been wasteful, we have at our disposal magnificent undeveloped resources that could provide our people with the means of remaining strong in peace and war, and peace forever. I propose that we proceed in the American way to develop our resources intelligently and without waste, and to give the whole population access to this . . . wealth."[12] In this unpublished article, "Rivers of Strength in War and Peace," Ickes argued for intelligent and efficient development by using the imperialistic discourse of empire building. Ickes was clearly mindful of environmental destruction, though this understanding was inflected by the need for economic progress. Ickes, perhaps more than most other Roosevelt administrators, embraced a utilitarian perspective that recognized tradeoffs where others saw only positive outcomes. Ickes alone pointed out the (necessary) harm that is done to nature. He justified this with the 'greatest good' argument, and used this frank discussion of tradeoffs to buttress his argument for empire building and high modernist state interference with nature.

These ideas were further elaborated in his December 1937 statement at the opening for bids on the second phase of Grand Coulee construction, the building of the high dam on the foundation.[13]

> Today we are taking the second step in the building of the greatest dam in the world. The base of Grand Coulee Dam, already the most massive masonry structure built by man, is nearly finished in the bottom of the Columbia River canyon, west of Spokane. Now bids are being opened for the completion of this tremendous structure to its full height of 553 feet. The work already done and that covered by the new contract will involve the placement of three times as much masonry as went into the Great Pyramid of Cheops. Here is a job worthy of the best that science and engineering, that labor and industry have to offer. Grand Coulee Dam will endure for many thousands of years; testifying to the imagination, the ingenuity and the skill of this generation.

> However, if I may be permitted to venture a prophecy, it is that in distant years to come this project will have given us a more coveted title than simply that of 'builders of great dams.' In saying this I do not mean to detract from the honor due to the engineers and the workmen who are building Grand Coulee Dam. Even they, I think, will agree that the significance of the dam is not found alone in the magnitude of its dimensions nor in the workmanship that has gone into its construction. It lies rather in the ends which are to be served.

Ickes argued that Grand Coulee will "have a profound influence on the social and economic life of the Pacific Northwest, which will be reflected throughout the country." Furthermore, "all of this will mean new homes and new opportunities for half a million or more Americans and new wealth for the Nation." Ickes used the rhetoric of empire building, economic progress, and the "best that science has to offer" to justify and celebrate the building of Grand Coulee Dam.

Harold Ickes was more than just FDR's close advisor. He was also the Secretary of the Department of Interior, a government agency that was beginning to recognize the importance of public opinion. In May of 1941 folk singer and songwriter Woody Guthrie was hired to write songs about the Reclamation projects in the Pacific Northwest (Guthrie 1943).

Guthrie, an Oklahoma native who followed the other farmers, workers, and drifters west during the Great Depression, was hired (at the recommendation of music historian Alan Lomax) by Sam Kahn, the Information Assistant for the Bonneville Power Administration (BPA), the government corporation that sold power from Bonneville and Grand Coulee Dams. Kahn and filmmaker Gunther Von Fritsch worked together to produce a series of documentary films about the BPA and Columbia Basin projects, the first of which was the film *Hydro*. The second was to be called *The Columbia*, about the Columbia River and its dams, but the project was scrapped with the U.S. entry into World War II. Hired under the title of "Information Consultant" to the BPA, Guthrie had an astonishingly prolific month of employment.[14] Guthrie produced 26 songs in the 30 days of his government job, making it the most productive month of his career (Murlin 1991: 10–11). The songs took the Columbia River as their main subject, with the Bonneville and Grand Coulee Dams as central objects of honor.

The hiring of Guthrie allowed the Department of Interior to spread public confidence by distributing potentially popular songs about the dams. Hiring Guthrie in particular helped to avoid resistance from the political left. Guthrie was an inveterate protestor, a socialist, a union supporter, and an itinerant activist. Guthrie's fame among the political left in the U.S. was well-established by 1941 (Klein 1980: 111–35), but his work was yet to be generally popular among most Americans (his famous song "This Land Is Your Land" would not become a familiar American ballad until near his death in 1967, in fact, the author remains somewhat anonymous even to those who know the lyrics by heart).[15] His support of the Grand Coulee project could show other socialists what Guthrie believed, in Joe Klein's words (1980: 195): "That the government was building all these massive dams was especially thrilling. It was what socialism would be like when it came to the United States."

In the song "Grand Coulee Dam" Guthrie played upon the imperialistic discourse that was popularized in the press by Robert Ormond Case (1935):

> Oh the world is seven wonders
> So the travelers always tell;
> Some gardens and some towers,
> I guess you know them well.
> But now, the greatest wonder
> Is in Uncle Sam's fair land;
> It's the King Columbia River
> And the big Grand Coulee Dam.

Guthrie compared the seven world wonders with Grand Coulee Dam, and found it to be the next greatest achievement. Furthermore, the dam was not just on *any* river. Guthrie focuses the discourse of empire building by labeling the river that the dam stops as the "*King* Columbia River."

Guthrie's rhetoric tends towards the grandiose:

> In the misty crystal glitter
> Of the wild and windward spray;
> Men have fought the pounding waters
> And met a watery grave.
> Well, she tore their boats to splinters
> But she gave men dreams to dream
> Of the day the Coulee Dam would cross
> That wild and wasted stream.

Building on the Hebraic discourse, he called the river "wild" and described its "pounding waters" where brave pioneers who dreamed of harnessing the wasted river have met their "watery graves." Guthrie thus combined the imperialistic and Hebraic discourses in telling the story of how the U.S. "took up the challenge" of putting the river to work for "the farmer and the factory and all the you and me." The result of this heroic work was to make the "factories hum" and to produce "Flying Fortress" airplanes that could "fight for Uncle Sam."

Uncle Sam took up the challenge
In the year of '33.
For the farmer and the factory
And all the you and me.
He said, "Roll along Columbia,
You can ramble to the sea.
But River, while you're rambling,
You can do some work for me!"

Now, in Washington and Oregon
You hear the factories hum'
Making chrome and making manganese
And light aluminum.
And there roars a Flying Fortress
Now to fight for Uncle Sam.
On that King Columbia
By the big Grand Coulee Dam.

The tune "Roll on Columbia" (a portion of which makes the epigram for this chapter) continued this combined rhetoric about chosen pioneers and imperialism.

(yes) Roll on, Columbia, roll on,
Roll on, Columbia, roll on.
Your power is turning the darkness to dawn
Roll on, Columbia, roll on.

Tom Jefferson's vision would not let him rest.
An empire he saw, the Pacific Northwest.
Sent Lewis and Clark, and they did the rest.
It's roll on, Columbia, roll on.

And far up the river is Grand Coulee Dam,
The mightiest thing ever built by man,
To run the great factories and water the land,
It's roll on, Columbia, roll on.

Guthrie uses the history of westward expansion—Jefferson's Louisiana Purchase and Lewis and Clark's expedition to map the new region—to evoke the empire that America was destined to become, when power from the Grand Coulee Dam would help turn "darkness to dawn."

In "Talking Columbia," however, Guthrie moved away from the two discourses of imperialism and the chosen pioneers, and brought an ironic tone to his lyrics:

You just watch this river and pretty soon
everybody's gonna be changing their tune.
The big Grand Coulee and the Bonneville dam
are running a thousand factories for Uncle Sam
and everybody else in the world
Making everything from sewing machines
to fertilizer, atomic bedrooms . . .
Plastic! Everything's gonna be made out of plastic.

Yeah, Uncle Sam needs wool, Uncle Sam needs wheat
Uncle Sam needs houses and stuff to eat.
Uncle Sam needs water and power dams.
Uncle Sam needs people, and people need land.
Don't like dictators none much myself,
but I think the whole country should be run by
Electricity!

On the face of it, Guthrie supported the goals of FDR and the New Deal (Uncle Sam needs water, power, and people, while people need land to live on and farm). But Guthrie earlier made an ironic turn, claiming that everything would be made out of plastic and run on electric power. Though he may have been taking the opportunity to make a joke, the sarcastic tone in some of his and Jack Elliot's recordings of this tune suggests skepticism of the governmental development plans.

Hiring Woody Guthrie may have been a simple and cheap public relations coup for the federal government, but it also helps to explain the lack of contestation in the discourse around the Grand Coulee. The success of the discourse of imperialism and the Hebraic discourse of a chosen people and a chosen land both hinged upon popular support. By hiring Guthrie to laud the dams in song, the government was further able to corral this support. But Guthrie was more than just a popular songwriter—he was a well-known protestor among both exodusters and New York socialist intellectuals. His songs about the Dust Bowl were generally critical of the government and the elites of the nation. His songs continually bashed the bankers who foreclosed on poor farmer's properties, forcing them to travel west; the rustlers and criminals of the West were compared favorably to the bankers and bureaucrats. In the song "Pretty Boy Floyd," Guthrie has the outlaw Floyd remark:

You say that I'm an outlaw,
  you say that I'm a thief
Now here's some Christmas dinner
  for the families on relief.

Now as through this world I rambled
I've seen lots of funny men,
Some will rob you with a six-gun,
and some a fountain pen.
Now as through this life you travel,
and through this life you roam
You won't never see an outlaw
drive a family from their home.

While the lyrics of "Pretty Boy Floyd" paint a negative picture of the elites of the country relative to the "outlaws," Guthrie saved his most biting sarcasm for the members of the clergy who had seemingly forgotten the core of social justice in the Christian faith as Guthrie understood it. Guthrie had a deeply-held ambivalence that was expressed in his music. While on one side he saw the corruption of many bankers, church officials, and government bureaucrats, he also saw the benefits that large public works projects could provide jobs for needy workers. Guthrie himself was struggling to provide for his family in 1941 when he was hired to write the songs about developing the Columbia River. This ambivalence was starkly exhibited in the tune "This Land Is Your Land."

Guthrie's classic song, written in 1940 and first recorded in 1944, described a patriotic vision of America, with "wheat fields waving," "diamond deserts," and "golden valleys." These patriotic symbols, however, are contrasted with the "dust clouds rolling." Guthrie's ambivalence, however, goes deeper than simply contrasting imagery.

The traditional song that most Americans know so well started life as a much darker picture. The song is a parody of Irving Berlin's "God Bless America," which Guthrie rewrote as "God Blessed America" (Klein 1980: 140). In addition to the lyrics that ended up in the recorded version of "This Land Is Your Land," in the original, Guthrie penned the following lines:

I roamed and rambled and followed my footsteps
To the sparkling sands of her diamond deserts
And all around me, a voice was sounding:
God Blessed America for me.

Was a big high wall there that tried to stop me
A sign was painted said: Private Property.
But on the back side, it didn't say nothing
God Blessed America for me.

One bright shiny morning in the shadow of the steeple
By the relief office I saw my people

As they stood there hungry,
I stood there wondering if
God Blessed America for me.

In this version of the song, the church and the relief office are juxtaposed, showing the inability of either to solve the hunger of the American people. In Guthrie's eyes neither the officials in Washington, nor in local houses of government, nor those in the churches were competent. Furthermore, the specter of private property is intimated as the source of the problem: high walls and signs that keep people out lead directly to hungry folks standing in line. Guthrie represents the observer who can cross the walls and see the other side of the signs, and though Guthrie doesn't spell it out for his audience, transcending private property and disobeying warnings would seem to be part of an answer to the nation's problems.

As a social critic, Guthrie's participation in praising the dams of the West therefore takes on a great deal of meaning. As Klein points out in his biography of Guthrie, in the early 1960s many of Guthrie's old friends felt that the radical message of his music was being diluted. In 1966 when the BPA named a power substation after Guthrie, his old associate Irwin Silber commented "They're taking a revolutionary, and turning him into a conservationist" (Klein 1980: 434). Though Guthrie's decision to write the songs about the dam was probably a straightforward one (he was broke and needed to feed a family), this governmental support was the basis for Guthrie's most productive composition period. High modernism has an ability to subsume contestation, and here we see quite clearly how Guthrie's ambivalence is (perhaps unintentionally) used by federal officials to further the state's development agenda. One further imagines that if other protestors on the political left saw Woody Guthrie supporting the dam, would they think twice about vocally opposing the project.

CONCLUSION: ROLL ON COLUMBIA

The discourses around Grand Coulee Dam are important in three respects. First, the lack of a powerful oppositional discourse precluded the possibility that the dam would not be built. Second, the discourses expose a set of constructions of nature and the nature-society relationship. Third, the idea of a dam was kept alive by a vibrant civil society that pushed the government to build an infrastructure for economic and political growth in the West.

Discourses over Grand Coulee largely fall into two categories: imperialistic and Hebraic. Different groups used these two categories differently: the national-level discussions were centered around a fairly pure imperialistic discourse, while the local proponents fought over specific proposals for the dam using differing styles of imperialistic high modernist rhetoric and individualistic capitalist rhetoric. Both of these local imperialist styles were cross-cut by a Hebraic discourse that portrayed the region as a chosen land. While the local elite groups used an individualistic capitalist discourse that was suffused with the values of unfettered competition and a strong opposition to state intervention, the proponents of the high dam argued from a high-modernist perspective that valued the interventions of the state in building large-scale water systems that could otherwise not be achieved by local capital. Both of these sides used Hebraic and imperialist discourses in addition to the individualist or high-modernist rhetoric.

The discourses of the two main adversarial groups precluded any discussion of not building a dam. The competition between discourses was instead over who would build the dam and what it would look like: Not only did the state-sponsored, high modernist plan win approval, but those who were opposed to a state-sponsored plan eventually backed it in order to get their part of the benefits. Even those one might expect to protest the dam ended up supported it. Local Indian tribes, whose land was inundated by the dam's reservoir, supported the dam due to the water and hydropower benefits. Farmers, who faced competition if new lands were brought into production, supported the dam due to its cheap hydropower. Labor unions, a group that often opposed non-unionized public works, supported the project because of the need for jobs. Woody Guthrie—opponent of government, church, and capitalism—supported the dam for all of the reasons listed above. High modernism as an ideology is largely successful because it can absorb resistance and suppress dissent.

As well as exposing the ability of high modernism to absorb resistance, the imperialistic and Hebraic discourses also expose a particular construction of a nature-society relationship. The rhetorical strategies employed by Woods, O'Sullivan, and so many others portray nature in complex ways, but ones that always revealed an attitude of domination on the part of humans.

Woods declared that "Nature" was on the side of the pumping plan developers. Nature had provided the perfect location, and had even built a canal, seemingly just for the purposes of the human inhabitants of the area. This characterization of nature points to the complexity of the nature-society relationship as understood by many of the proponents of Grand Coulee Dam. The dam was at once intended by nature and yet the Columbia River was seen as a mighty force that was nearly impossible to subdue. Humans could harness the river, but it would take an awesome effort. Nature was clearly an active force in this schema: nature was capable of building canals, carving out a reservoir, and providing an ideal dam site. The gendered character of this relationship is inescapable: it is almost as if nature is inviting humans (men) to subdue it (her).

Yet nature was also seen as wild, powerful, and formidable as an opponent. Much of the imperialistic rhetoric is obsessed with describing the wild power of the river, albeit in terms of its potential for society. The river was characterized as the "wildest big stream in the civilized world," and the attempt to harness it would be impossible, "a waste of time an' money" (Woods 1937a). Thus nature was also an active force as well as one that invited humans to dominate it. What are we to make of this complexity and contradiction? In part, it stemmed from the contradictory character of the discourses used: imperialism implies domination—of nature and of other humans—while the Hebraic discourse implies that the land is blessed by god, and the inhabitants (or immigrants) are a chosen people. Thus nature is simultaneously a resistant force that must be overcome as well as a beneficial force that helps humans toward their glory.

Only Harold Ickes comes close to recognizing these complexities in the nature-society relationship. Ickes is a utilitarian at heart: there are certain tradeoffs that must take place to maximize positive benefits for as many Americans as possible. Yet the simple utilitarian dictum doesn't prepare one for Ickes' rhetoric, which shows a strikingly subtle analysis of how humans must develop resources wisely and carefully to protect nature from damage—some of which is caused by humans themselves. It is almost jarring, then, when Ickes concludes that humans must "interfere with nature in order to give her a chance to serve us." Nature is, ultimately, a resource for use by people for economic ends.

If the boosters of the dam manipulated interpretations of empirical facts regarding nature to make the dam seem blessed (or at least easy to build), then what does it matter whether nature is a positive or a negative force, or whether imperialistic and Hebraic discourses are contradictory? This is to say, in the end the dam is built, and all of the competing social groups come out ahead. If so, why do the competing discourses matter? These competing discourses were all self-interested. The imperialistic and Hebraic discourses were both heavily disposed towards a society-centered philosophy. In fact, there were no oppositional discourses that were not self-interested until the rise of the environmentalist discourse during the Glen Canyon debate. Until this powerful environmentalist discourse emerged, there is no apparent opposition, or at least no discursive grounds to root opposition in. Lacking any effective oppositional discourse, the ideology of high modernism is hegemonic.

Chapter Six

Glen Canyon: The Last of the High Dams

Yesterday the Colorado River was a natural menace. Unharnessed it tore through deserts, flooded fields, and ravaged villages. It drained the water from the mountains and plains, rushed it through sun-baked thirsty lands, and dumped it into the Pacific Ocean—a treasure lost forever.

Today this mighty river is recognized as a national resource. It is a life-giver, a power producer, a great constructive force. Although only partly harnessed by Boulder Dam and other ingenious structures, the Colorado River is doing a gigantic job. Its water is providing opportunities for many new homes and for the growing of new crops that help to feed this nation and the world. Its power is lighting homes and cities and turning the wheels of industry. Its destructive floods are being reduced. Its muddy waters are being cleared for irrigation and other uses.

Tomorrow the Colorado River will be utilized to the very last drop. Its water will convert thousands of additional acres of sagebrush desert to flourishing farms and beautiful homes for servicemen, industrial workers, and native farmers who seek to build permanently in the West. . . . Here is a job so great in its possibilities that only a nation of free people have the vision to know that it can be done and that it must be done. The Colorado River is their heritage.

United States Bureau of Reclamation, 1946

Good men, who have plans for the Colorado River whereby "a natural menace becomes a natural resource," would argue tirelessly that the Colorado must be controlled, that its energy should be tapped, and sold to finance agricultural development in the arid West. But our point here is that for all their good intentions these men had far too insular a notion of what man's relation to his environment should be, and it is tragic that their insularity was heeded. The natural Colorado—what is left of it—is a miracle, not a menace.

David Brower, 1963

The imperialistic high modernist discourse that was so well characterized by the proposal and construction of Boulder and Grand Coulee Dams, maintained its hegemony through the 1940s and into the 1950s. As the epigram from the Bureau of Reclamation above shows, the well-worn rhetoric of state building, national importance, and the other strands of the imperial high modernist discourse were recycled to great effect as dams were put up on almost all of the wild streams of the arid West. But in the early 1950s the rise of an environmentalist discourse allowed a successful contestation of two dams in Utah. This newly revived discourse about the human relationship with nature focused on the aesthetic qualities of the environment, and argued the need for conservation and preservation. In the words of David Brower, reproduced in the second epigram above, the Colorado River was a "miracle, not a menace."

The discourse around Glen Canyon Dam is marked by the rise of this contestation by individuals and social groups that were not simply self-interested. This oppositional discourse highlighted the inherent aesthetic value of natural and wild areas over a strict economic logic. The generally middle-class groups who crafted this revived discourse were interested in preserving areas as they existed, without development, roads, or other human artifacts. This preservation was, in many ways, self-interested. Many of the proponents of preservation were outdoor-recreation enthusiasts who wanted to enjoy the area in an unsullied (from their perspective) form. As discussed in chapter two, there was a class basis to the rise of the environmental movement in the 1950s, where middle class groups organized around the increased visitation to national parks and other natural areas by middle class vacationers. While these groups were clearly self-interested because they wanted to protect nature for human aesthetic consumption, they were also interested in preserving the areas both for broad ecological reasons and in the interest of future generations.

The discourse is thus focused on nature as an aesthetic and spiritual resource that can benefit humans in important ways outside of economics. While the imperialist modernist discourse was heavily disposed towards a society-centered philosophy that regarded nature as having only economic value, the environmental discourse at Echo Park and Glen Canyon was focused on nature as having other forms of value. Some of these forms of value were similarly society-centered, such as the supposed ability of nature to provide respite from urban landscapes and its restorative powers for dehumanized workers and other city-dwellers. In this way, the mid-twentieth century middle class environmental consciousness viewed nature as an aesthetic rather than an economic resource, but a resource for consumption none-the-less. As Cronon (1996) has argued, part of the trouble with wilderness is that we have defined it as exclusive of humans. Wilderness is an area that has no marks of human passage, has not contained humans—it is a radical Other for human society. Such an area, as discussed in chapter two, is quite hard to come by. Most of the areas that environmentalists wanted to preserve had been home to indigenous peoples who had been forcibly removed so that Anglos could log, mine, and otherwise make use of what they called the frontier. Views of nature that exclude humans, then, are at best historically inaccurate. But to leaders of the Sierra Club, the Isaac Walton League, and the Wilderness Society in the mid-twentieth century, wilderness was to be preserved for human appreciation—for aesthetic consumption rather than economic consumption. But these groups also expressed other ideas about the value of nature, including ecological and environmental values that did not relate directly to humans. This powerful discourse grounded the opposition to the ideology of imperialistic modernism.

Similar to Boulder and contrary to Grand Coulee, the debates at Glen Canyon and Echo Park involved a strong voice on the part of the Bureau of Reclamation. This voice characterizes a discourse of imperialistic modernism very similar to the early proposals around Boulder Dam. Contrary to the case of Boulder and similar to Grand Coulee, the role of civil society was overpowering. In the case of Glen Canyon, however, one of the voices of civil society contested the government plan for development. This oppositional discourse is best labeled *environmentalist,* as it attempted to refocus society on the aesthetic and ecological values of nature, rather than on the economic logic of nature as a natural resource for human consumption. This chapter investigates the rise of this effective oppositional discourse in the 1950s, centered on responses by groups in the public sphere to the Bureau's proposal to dam Echo Park and Glen Canyon. I argue that this period marked the end of the "golden years" of dam building, and that the Glen Canyon case represents a significant change in the relationship between society and nature. This change is marked by the rise of contestation around Glen Canyon Dam, but its emphasis is less on the individual dam than on advocating a shift from a nature-society relationship based primarily on domination and economic-resource maximization to one based in part upon aesthetic and other forms of appreciation.

The discourses at Glen Canyon thus fall into two main oppositional categories. The Bureau of Reclamation was the leading proponent of building the dam along with the many other dams and canals in their 1946 upper Colorado River Basin development plan, using the typical arguments of imperial modernism. Opposing this development in general, and specific dams in particular, were environmental groups led by the Sierra Club and the Wilderness Society. These groups organized resistance to the Bureau's plan and helped to define a discourse of opposition based in an appreciation of the environment over development. To these two main discourses was added a third after the completion of Glen Canyon Dam. This third discourse is one of cooptation, whereby the Bureau of Reclamation responded to the oppositional rhetoric by emphasizing the aesthetic quality of the reservoir. This discourse of *improved nature* argues that what is created by humans is actually better than what existed before.

AN INTRODUCTION TO THE SOCIAL, POLITICAL, AN GEOGRAPHIC HISTORY OF GLEN CANYON DAM

Glen Canyon Dam, constructed between 1956 and 1964, is on the upper Colorado River, between Arizona and Utah, south of Bryce Canyon National Park, and at the northern border of the Navajo Nation Reservation. The arched concrete dam was built 710 feet high; its crest height is precisely that of Boulder Dam, 583 feet from the original river channel to the roadway atop the dam. Glen Canyon was named by John Wesley Powell on his journey down the Colorado in 1869; Powell was struck by the deep and cool canyon that supported a lush vegetative ecosystem so different from the rest of the desert region. These damp green canyons reminded Powell of glens in his native Scotland, so he christened the site Glen Canyon. After a successful career as a Union Captain in the Civil War, where he lost an arm, Powell, with governmental backing, embarked on a mission to explore the lands and rivers of the Southwest. He and his crew of ten men were the first Anglos, and perhaps the first humans, to navigate and map the Colorado River and several of its tributaries. That they did so in four wooden boats, and Powell with only one arm, earned them a hero's welcome in the western territories. Powell went on to become the first Director of the United States Geological Survey upon its creation in 1879, and advocated strongly for watershed-wide governance, and reduction of acreage restrictions for western ranching (Alexander 1960; Martin 1989; Powell 1878; Reisner 1986; Stegner 1954).

Glen Canyon was first proposed as a site in 1921 as a response to Phil Swing and A. P. Davis's proposal for a dam at Boulder Canyon. Opposition to Boulder Dam (see chapter four) was in part headed by politicians in Arizona, who saw that a dam on the upper Colorado River served the interests of their state better than the Boulder Canyon site. Arizona lost this battle, but the Bureau of Reclamation kept detailed surveys of the Glen

Canyon site, and in 1946, it proposed a sweeping upper Colorado River Basin development that included Glen Canyon. Glen Canyon Dam was to be part of a string of large storage dams on the upper Colorado and its tributaries. This comprehensive development plan also included dams at Echo Park and Split Mountain on the Green and Yampa Rivers. The dams would have backed water into Dinosaur National Monument in northeast Utah.[1] Opposition to this encroachment into the National Park System eventually stopped the development of these dams, the first effective opposition to a large-scale water development in the arid West.

Through negotiation with environmental groups, the Bureau of Reclamation agreed to enlarge Glen Canyon Dam instead of building those that would encroach upon Dinosaur National Monument. Glen Canyon was dammed, but the Sierra Club and other national environmental organizations mourned its loss and vowed never to let this happen again. So far, they have been successful.

HARNESSING THE COLORADO: THE BUREAU'S GRAND PLAN

In the late 1940s and early 1950s the Bureau of Reclamation built dams at an incredible pace throughout the West. Rivers by the hundreds were dammed for single and multiple purposes as the Bureau rode a wave of public and governmental approval. This golden age of dam building was overseen by Commissioners of Reclamation Michael Strauss and Floyd Dominy, who pursued further construction with great zeal. One major aim of the Bureau was the total development of the upper and lower Colorado.

In 1946 the Bureau of Reclamation published its plans for this total development of the Colorado River. The ponderous title of the document conveys the enormity of its contents: *The Colorado River: A Comprehensive Report on the Development of the Water Resources of the Colorado River Basin for Irrigation, Power Production, and Other Beneficial Uses in Arizona, California, Colorado, Nevada, New Mexico, Utah, and Wyoming* (Bureau of Reclamation 1946). This report reuses the label given to the Colorado 25 years before when Boulder Dam was proposed, the epigram printed on the cover of the report reads: "A natural menace becomes a national resource." This continuity in discourse is important. The Bureau, with its comprehensive report, was attempting to continue its successful development of the Colorado and the West through what had become standard rhetorical techniques. The discourses used at Boulder and Grand Coulee were unquestioningly re-used to boost the Glen Canyon and Echo Park dams in the upper Colorado Basin project.

The report outlined a total of 134 projects (including dams, canals, diversions, and pumping systems) in the upper and lower Colorado Basin, totaling $2,185,442,000. Included in these were proposals for dams at Echo Park, Split Mountain, and Glen Canyon. In the report, the Bureau outlines the justifications for such a massive series of projects:

> Future development of the water resources of the Colorado River Basin is needed to relieve economic distress in local areas, to stabilize highly developed agricultural areas, and to create opportunities for agricultural and industrial growth and expansion throughout the Colorado River Basin (Bureau of Reclamation 1946: 21).

The Bureau used a typical rhetoric of economic progress to justify its proposals. It emphasized the industrial and agricultural growth that would be spurred by the projects, as well as the relief of local economic distress. Such arguments had become, by the late 1940s, standard techniques for legitimation. The Bureau's new projects, however, would be both bigger and, it argued, more beneficial.

For example, in their 1946 proposal, the Bureau claimed that the cost to benefit ratio was higher than 1.0. "These benefits indicate that a basin-wide plan for full development of the water resources could return to the Nation $1.30 for each dollar required to construct, maintain, and operate the projects" (*ibid.* 1946: 18).

And yet, the Bureau did identify some cause for hesitation. Through a careful reading of the document it is clear that the Bureau admits "there is not enough water available in the Colorado River system for full expansion of existing and authorized projects and for all potential projects outlined in the report" (*ibid.* 1946: 21). So why did the Bureau propose them if there was not enough water?

The answer may be that the Bureau saw the Colorado River as teetering on the brink. With just a small ($2 billion) nudge, the Bureau could knock the Colorado into the realm of completely harnessed. The Bureau argued that:

> Yesterday the Colorado River was a natural menace. Unharnessed it tore through deserts, flooded fields, and ravaged villages. It drained the water from the mountains and plains, rushed it through sun-baked thirsty lands, and dumped it into the Pacific Ocean—a treasure lost forever. Man was on the defensive. He sat helplessly by to watch the Colorado River waste itself, or attempted in vain to halt its destruction (*ibid.* 1946: 25)

The Bureau here was engaged in the discursive construction of the river. The river was simultaneously a powerful actor ("a natural menace") and also an entity that was treasured as a potential economic resource. The justification based on an imperial modernist ideology of expansion and development by the federal state for the utilitarian benefit of all society was founded in such a rhetorical construction. It is no mistake that only "villages" populate the area through which this unharnessed resource travels. The state was interested in building up civilization in these areas—never mind that Los Angeles, San Diego, and Phoenix were already sprawling metropolises at the time.

Man was portrayed as defensive against the active river, but through courageous action, the river could be tamed:

> Today this mighty river is recognized as a national resource. It is a life-giver, a power producer, a great constructive force. Although only partly harnessed by Boulder Dam and other ingenious structures, the Colorado River is doing a gigantic job. Its water is providing opportunities for many new homes and for the growing of new crops that help to feed this nation and the world. Its power is lighting homes and cities and turning the wheels of industry. Its destructive floods are being reduced. Its muddy waters are being cleared for irrigation and other uses (*ibid.* 1946: 25).

The river had been tamed and transformed through the beneficent hand of the state. The Colorado now gave life rather than taking it. It had been put to work, had built new homes, gardens, and fields, contributing to national prosperity. The proper role of the river had been achieved, with a little help from humankind. And yet the job was not done, according to the Bureau.

The river was only partially harnessed. Given the terrific benefits gained already, what a shining future the river had before it:

> Tomorrow the Colorado River will be utilized to the very last drop. Its water will convert thousands of additional acres of sagebrush desert to flourishing farms and beautiful homes for servicemen, industrial workers, and native farmers who seek to build permanently in the West. Its terrifying energy will be harnessed completely to do an even bigger job in building bulwarks for peace. Here is a job so great in its possibilities that only a nation of free people have the vision to know that it can be done and that it must be done. The Colorado River is their heritage (*ibid.* 1946: 25).

In this amazing nationalistic passage, the Bureau claimed legitimacy for its state-building proposals through the great prosperity to be gained from further development. The 134 dams and canals in its proposal were labeled as the heritage of past Americans, those free frontiersmen who worked so hard to build upon the vision of manifest destiny. The Colorado River remained a "part of America's frontier" (*ibid.* 1946: 71), the inheritance of all Americans, whom the Bureau glowingly called "empire builders" (*ibid.* 1946: 45).

The Bureau's grand plan had many supporters; most prominent was the State of Arizona. Officials from Arizona used the same discourse of imperial modernism to boost Glen Canyon Dam, arguing that although the dam was long overdue, it was required to bring development to their state. In April 1957, for example, *Desert Magazine* described the benefits to arise from Glen Canyon Dam, using similar rhetoric as the Bureau (Murbarger 1957):

> When man erects a mighty dam across the Colorado River at Glen Canyon . . . a new era will dawn. . . . A city will rise from the desert floor; new factories will turn their wheels with power from the impounded water.

The building of the dam was hailed as the start of a new era, one filled with prosperity for the population and industry. The article further boosted the dam with discussions of the benefits of recreation on the reservoir and the huge areas of shoreline that would be created by the lake for tourists to explore.

Many groups of Arizonans boosted the dam in newspapers and periodicals. Thomas Ellinwood, a political cartoonist who had lampooned California's dominance in the West during the debates over Boulder Dam also celebrated the building of Glen Canyon. The Phoenix periodical *Action* published an article in its October 1957 issue boosting the dam, arguing that the long range benefits for Arizona were clear:

> No doubt about it, northern Arizona, particularly Flagstaff, will benefit from the building of the Glen Canyon Dam. Phoenix, because of its strategic position in the state's economic pie, will also benefit.

Arizona boosters focused on the economic benefits that the dam would produce, combined in part with flood control. The discourse they used to do so was strikingly similar to that of the imperial modernist discourse used at Boulder and Grand Coulee. Echoing the Bureau of Reclamation's recycling of a successful discourse, the Arizona supporters discursively constructed the river as an economic resource that was finally being developed so as to bring Arizona what was its due.

In what appears to be an attempt to counter protests about the building of the dam, some periodicals engaged in discourse that constructed the area as a wasteland. In February of 1957, the magazine *Western Construction* argued that:

> The entire area is a vast expanse of wasteland, uninhabited except for a few ranchers on the northwest side of the river and scattered Indian families on a reservation to the southeast (McClellan 1957: 29).

In fact, the Navajo Nation Reservation (the largest in the U. S.) had thousands of inhabitants and one of the most developed rangeland economies of any reservation. The construction of the area as one that could easily be sacrificed can be seen as a response to the environmentalist discourse highlighted in the next section.

The Bureau of Reclamation and its allies in civil society recycled many of the rhetorical strategies that were successful in the 1920s and 1930s for boosters of the Boulder and Grand Coulee Dams. This state-building discourse combines legitimation through the interpretation of history (frontiersmen of vision who built a foundation for the future) with the rational application of science and technology to benefit all society (reduction of a flood menace, improvement of an economic resource). This combination allowed the discursive reconstruction of the Colorado River into a natural economic resource. The river was tamed and harnessed and put to work for society.

The only problem with this discourse was its overuse; the Bureau could have had no idea that this was reducing the efficacy of the discourse, for it had been so successful in the past. A new way of thinking about nature was growing in the American West, however, and its rise eclipsed the Bureau's dominance in western development.

CHANTING DOWN ECHO PARK

The environmentalist discourse used to oppose Echo Park and Split Mountain Dams in the 1950s was not actually new. It was a derivation of the nature-as-aesthetic-resource argument that John Muir unsuccessfully used from 1907 to 1913 in fighting the Hetch Hetchy Dam in California. In fact, elements of its expression can be seen in works that date back at least 120 years (Nash 1967/1982). John Muir was one of the first advocates of wilderness preservation in the Sierra Nevada Mountains of California and Nevada. His founding of the Sierra Club in 1892 was partially in response to the conversion of Yosemite State Park into the second National Park. As stated in a 1911 bulletin, the Club's goals were primarily to "take the lead in all matters involving the preservation of the wonderful natural scenery which California is so fortunate as to possess, and in calling the attention of the world to these wonders" (Sierra Club 1911).[2]

In 1908, these goals were threatened by the proposal of a dam in Hetch Hetchy Valley, adjacent to Yosemite Valley and partially within the Park boundaries. Muir led the battle to save Hetch Hetchy Valley, arguing in a 1908 letter to Sierra Club members that Yosemite and Hetch Hetchy held an "unrivalled aggregation of scenic features" and that it should be "preserved in pure wilderness for all time for the benefit of the entire nation"(Muir 1908). Muir argued that the government should respect the boundaries of Yosemite National Park, or else all such boundaries would be meaningless. In the end, Muir's battle was lost, and San Francisco built a dam for its municipal water supply in Hetch Hetchy Valley. But this oppositional discourse was resurrected 40 years later by the very group that Muir had founded—a discourse that placed inherent aesthetic value in nature.

The 1946 proposal by the Bureau of Reclamation to build a set of dams at Echo Park and Split Mountain, as part of the grand plan to develop the upper Colorado River Basin, would back water into Dinosaur National Monument. Bernard DeVoto broke this story of a latter-day Hetch Hetchy in the 22 July 1950 issue of the *Saturday Evening Post.* From his regular *Harper's* column "The Easy Chair," DeVoto had denounced cattle barons and Bureau of Land Management grazing leases (Thomas 2000). DeVoto's article "Shall We Let Them Ruin Our National Parks?" was a similarly incendiary piece, full of fighting energy and inflammatory rhetoric. Under the large-font title, the piece opens with a mid-sized-font sentence in offset text asking, "Do you want these wild splendors kept intact for your kids to see? Then watch out for the Army Engineers and the Bureau of Reclamation—because right where the scenery is, that's where they want to build dams" (DeVoto 1950:17).

DeVoto starts the article by challenging the democracy under which we ostensibly live:

> No one has asked the American people whether they want their sovereign rights, and those of their descendants, in their own publicly reserved beauty spots wiped out. Thirty Two million of them visited the National Parks in 1949. More will visit them this year. The attendance will keep on increasing as long as they are worth visiting, but a good many of them will not be worth visiting if engineers are let loose on them (*ibid.* 1950:17).

DeVoto warned Americans not to let the engineers of the Bureau perpetrate this crime against "unspoiled natural beauty," and continued with an admonition:

> No one can doubt that the public, if told all the facts and allowed to express its will, would vote to preserve the parks from any alteration now or in the future. But the public is not being told all the facts; it is not being given a chance to express its will (*ibid.* 1950:17).

DeVoto's muckraking article argues quite clearly that Americans would never choose to let this go forward, if only they knew. The piece is a bit disingenuous, for the Bureau engineers were not trying to pull the wool over anyone's eyes. They were in communication with the Park Service throughout the planning stages, and fully believed that the reservoirs represented a beneficial recreational opportunity for Americans. DeVoto, however, disagreed with this assessment. To DeVoto, the area was perfect as it was, and should not be altered in any way. If given all the facts, DeVoto argued, Americans would not support the project.

The Americans DeVoto speaks of were the readers of *Harper's,* supporters of the Sierra Club and other voluntary organizations, and many, many others. As discussed in chapter two, there was a class-basis for this movement that DeVoto tapped into. The rise of the middle class, with their suburban homes, newfound affluence, and most importantly, their cars, brought many Americans into contact with nature that they had not been exposed to before. This exposure, during vacations and weekend getaways as exemplified in Norman Rockwell's "Going and Coming," helped build support for protecting nature, as DeVoto argued Americans should.

Amidst half-page photographs of Dinosaur National Monument, DeVoto goes on to describe the scenic quality of the area as well as the ruin that it will become. Though he never explicitly compares the Dinosaur case to Hetch Hetchy, DeVoto's article proved quite significant to groups like the Sierra Club. The *Reader's Digest* reprinted the article later in 1950, and Martin Litton, reporter for the *Los Angeles Times,* wrote several articles exploring the case in more depth. It was this series of articles outlining the imposition on a wilderness area that caught the attention of David Brower, Executive Director of the Sierra Club (Litton 1992).

Brower took the lead in opposing the dam in Echo Park, seeing in the fight the possibility of redeeming Muir's loss at Hetch Hetchy nearly 40 years before. Brower assembled a coalition of individuals and groups committed to keeping national parks and monuments free from development. He led the fight by coordinating a massive letter-writing campaign and helping to publish many pamphlets and several books. Among the many notable figures involved in the fight were wilderness photographers Eliot Porter and Ansel Adams, novelist Wallace Stegner, and publisher Alfred A. Knopf.

Brower was very concerned about the encroachment of a reservoir into a national monument, and he recruited Wallace Stegner to edit a book on the Dinosaur situation, in an attempt to bring national attention to the cause, to be published by Alfred K. Knopf (Thomas 2000: 164). *This Is*

Dinosaur: Echo Park Country and Its Magic Rivers (1955) combined the writing talents of Stegner, Knopf, and several others with 36 striking photographs of the region, six of which took advantage of the new, and expensive, full-color printing technology (see Figure 6.1). The book's aim was to introduce people to this little-visited area, and to convince them that it was worth saving—a copy of the book was delivered to each member of Congress (Fox 1981: 285).

Wallace Stegner's contribution discussed the history of the national monument, an "almost 'unspoiled'" wilderness area. With his deep understanding of the intertwining of human history and natural environment, Stegner notes that Dinosaur National Monument is a "palimpsest of human history, speculation, rumor, fantasy, ambition, science, controversy, and conflicting plans for use, and these human records so condition our responses to the place that they contain a good part of Dinosaur's meaning" (*ibid.* 1955: 3).

In describing the area, Stegner talks lovingly of "cliffs and sculptured forms [that] are sometimes smooth, sometimes fantastically craggy, always massive" that "have a peculiar capacity to excite the imagination; the effect on the human spirit is neither numbing or awesome, but warm and infinitely peaceful" (*ibid.* 1955: 4).

Figure 6.1 Coffee Table Activism: Steamboat Rock, Dinosaur National Monument, frontispiece, *This is Dinosaur: Echo Park Country and Its Magic Rivers,* Wallace Stegner, Alfred A. Knopf, 1955. Reprinted by permission.

Stegner concludes by saying that the United States has entered an "Antibiotic Age"—meaning that it is against life. Stegner continued:

> We had better not be against life. That is the way to become as extinct as the dinosaurs. And if, as the population experts were guessing in November 1954, the human race will (other things being equal) have increased so much in the next three hundred years that we will have only a square yard apiece to stand on, then we may want to take turns running to some preserved place such as Dinosaur. . . . There isn't much [wilderness] left, and there is no more where the old open spaces came from (*ibid.* 1955: 14–15).

Stegner's call for wilderness protection came at a time when he was still relatively unknown as a writer. In 1955 his important early work *The Big Rock Candy Mountain* was ten years old, and he had just finished his nonfiction novel on John Wesley Powell's adventurous exploration of the Colorado, *Beyond the Hundredth Meridian.* His Pulitzer Prize-winning novel *Angle of Repose* would not be written until fifteen years later. Stegner was taking a chance by being politically outspoken. He was, as Thomas argues, in some ways attempting to fill the role left empty by the death of his friend Bernard DeVoto in 1955 (Thomas 2000: 166). Alfred A. Knopf was, in many ways, taking a bigger chance.

Knopf published *This Is Dinosaur* and presumably put forward much of the capital needed for the expensive camera work. His chapter, entitled "The National Park Idea," argues forcefully for preservation of wilderness areas for both people and wildlife. The national park "is not a resort, though there will always be those who try to make it so. And the very special purposes of recreation, education, refreshment, and inspiration for which Parks and Monuments have been set aside prohibit many economic uses which are thoroughly legitimate elsewhere" (*ibid.* 1955: 85). Knopf argued that there were many other places where such economic purposes could be pursued, but that National Parks and Monuments had to be defended categorically and on general principle, or else all such areas would be threatened in the future.

After arguing philosophically for the preservation of Dinosaur National Monument, Knopf argued from logic. Such a threat is not just temporary, but permanent:

> If you flood a canyon, as it is proposed to flood the Dinosaur canyons with dams at Echo Park and Split Mountain, that canyon is gone forever, buried first under water and eventually under silt (*ibid.* 1955: 86–87).

Much of the piece is spent arguing against perceptions of the American public about the southwestern lands (exemplified in articles such as the one previously quoted from *Western Construction*). Knopf is at pains to point out that

> Dinosaur is not expendable wasteland, not a profitless desert, but a scenic resource of incalculable value that has been preserved this long precisely because of its inaccessibility. . . . Dinosaur deserves to be more visited. . . . That is all it would take, that democratic groundswell, to insure that Dinosaur and the other superlative places will be passed on, unimpaired, to our grandchildren's grandchildren (*ibid.* 1955: 93).

Knopf ends by calling on the legitimacy of history and the myth of democratic America. Americans are wise people, and can see value when it is shown to them. They deserve their heritage, and so do their grandchildren. This treasure can be saved through the use of our democratic powers to stop the tyrannical exercise of authority by a faceless bureaucratic agency.

Knopf and Stegner's book enjoyed quite a good reception. It is styled as a coffee-table book, and was sold all over the country through Knopf's powerful distribution channels. In combination with the massive letter-writing campaign that Brower organized through the middle class members of the Sierra Club, the Wilderness Society, and the Isaac Walton League, the fight for Dinosaur ranged from American's living rooms to Capitol Hill.

Brower initiated a storm of protest over the proposed dam at Echo Park, a key element of which was the flooding of the Department of Interior and Congressional Representatives with letters of outrage. Letters were addressed to President Eisenhower, Secretary of Interior Douglas McKay, and individual Senators and Representatives. Most letters were forwarded to the Secretary of Interior, who cataloged many of them, now collected at the National Archives and Records Administration. The catalog for 1954 contained a listing of 2,875 letters that the department received that year alone. The letters are signed by individuals writing as rangers, lawyers, citizens, or members of conservation groups. All of these letters exhibit some form of an environmentalist discourse, often combined with an economic argument, such as evaporation problems, cost-benefit arguments, and even the perception that hydropower was obsolete due to the expansion of atomic energy (perhaps a very un-environmental argument).

The environmentalist discourse that the proposed dam at Echo Park brought out was focused on the quality of the place in and of itself. The construction of the river canyon as an economic resource was resisted vociferously. Instead, the river, the canyon, and the entire area were discursively constructed as a natural and aesthetic resource that was of such value for recreation and inspiration that to destroy it for economic purposes would be a great evil.[3] Letter writers sounded this tone over and over, in many creative ways.

This environmentalist discourse was straightforwardly exemplified by Edward Thatcher of Eugene, Oregon (8 March 1954), who cited the "magnificent natural beauty" of Dinosaur National Monument. Thatcher argued that the proposed Echo Park Dam would inundate the "canyon scenery and rock formations incomparable in their value to citizens of this country."[4]

Russell D. Butcher of Millbrook, New York spent a bit more time explicating his position. On 15 January 1955 Mr. Butcher wrote to the President, saying that he was "greatly disturbed" and that

> it is my belief that this country should protect its great parks from commercial and private developments. . . . I do not consider any one of these plans to be of great enough importance, or without perfectly good alternatives, to warrant a breaking away from park principles—of keeping them 'intact and in their entirety for the enjoyment, education, and inspiration of all the people for all time.' Also, because these park service areas comprise only about one-half of one per cent of the entire U.S., I truly believe that we should preserve them as a last remnant of the once vast primeval America. . . . It is merely opening the way for further encroachment upon other areas. It is very easy to visualize a slow eating away of the park system, as one by one they are opened up to commercial interests. I believe therefore, that we should start thinking about this problem now before we suddenly find ourselves without any of these fine parks; that we should pass them on to the next generations, unspoiled.

Vera Moran, of Santa Rosa, California, was not nearly as congenial and circumspect as many writers. Ms. Moran wrote to Secretary McKay saying:

> Those who want to benefit themselves by stealing public resources—whether forests, parks, national monuments-or however derogated—are
>
> Public Enemies
> Of the United States
> When they get through with it, America the Beautiful will no longer be beautiful—it will be stripped and stark.
> . . . protect the public and public interests by saying to these predators and public enemies:
> Keep Out!!

Such arguments about the splendid beauty of Dinosaur were clearly heart-felt. Many writers went even further in their claims about the uniqueness of the area. A. Weston Niemela, of the Chemketan Conservation

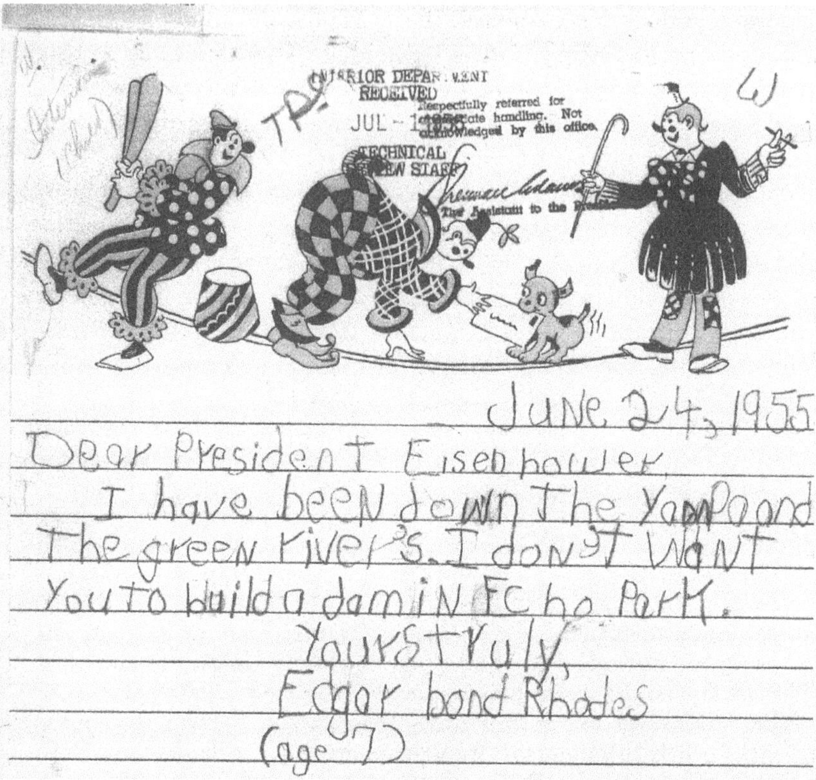

June 24, 1955

Dear President Eisenhower,

I have been down The Yampa and The green river's. I don't want You to build a dam iN Echo Park.

Yours truly,

Edgar bond Rhodes

(age 7)

Figure 6.2 A child's letter to President Eisenhower, 24 June 1955.

Committee, a conservation group within the Oregon Indian tribe, wrote on 5 March 1955 that "Many of us in the Chemeketans have been to the Monument and can testify to its unique beauty; as an area of recreational and spiritual value it could never be replaced." In the nuanced version of the Chemeketans' discourse, the uniqueness of this area in terms of its beauty and recreational qualities is combined with a spiritual element. The spiritual aspect of their discourse makes a larger set of claims that evokes John Muir's idea of nature as a cathedral for worship: inundation equals desecration.

On 28 April 1955, Eleanor Roosevelt Elkott, of Birmingham, Michigan, wrote to the President, saying "The United States is a big country. The citizens derive spiritual and moral strength from their land—touring, camping, fishing, golfing. It is not fair for citizens who believe in freedom and democracy to be overuled [*sic*] by men sitting in offices who want to make money. . . . We must not build Echo Park Dam." Her association of golfing with spirituality not withstanding, Ms. Elkott makes a case that was echoed by many writers.

The popular environmentalist discourse tended to cite the spiritual, recreational, and inspirational qualities of wilderness in general and the American Southwest in particular. The letter writers intensely resisted the construction of the Monument as a wasteland or as an economic resource. Instead, they saw the canyons and the rivers as an incomparable aesthetic resource that should not be squandered in the name of progress and economic development.

Many writers who opposed the dams added an economic logic to their arguments about the aesthetic value of the area. Such letters, however, often had a more *ad hoc* character to them. This suggests that the cost-benefit portion of these letters was individually crafted, while much of the environmentalist arguments were based largely on the national discourse of groups such as the Sierra Club. The letters tended to combine a concern for preservation with the idea that such a hydropower project was either unnecessary or that the benefits seem much reduced from what the Bureau claimed. Dan Blackly of Maryland, for example, wrote on 23 March 1955 that "just as we are on the verge of obtaining unlimited power from the atom it would seem foolish to destroy Dinosaur National Monument for this purpose."

Porter Butts, of Madison, Wisconsin, wrote to the President on 21 March 1955. Mr. Butts argued that many government agencies oppose the project (presumably meaning the National Park Service), and that even the Bureau of Reclamation's own reports have been shown to be erroneous. "I appeal to you to save this spectacular, irreplaceable area that belongs to all of us, especially since it has been shown that the water and power needs of Colorado and Utah can be met without sacrificing one of the unique natural marvels of our country." Mr. Butts connects two strands of argument regarding the Monument: an environmentalist and an economic argument.

J.C. Bradley, also of Madison, Wisconsin, wrote to President Eisenhower on 29 November 1955, reminding him of his recent statement that the Tennessee Valley Authority represented "creeping socialism." Why, Mr. Bradley asked, did the President then support the dam at Echo Park?

Other writers, such as John Ripley Forbes, of Sacramento, California (3 February 1954) called on President Eisenhower to act like the Roosevelts:

> As good a republican as Theodore Roosevelt and as good a democrat as Franklin D. Roosevelt fought throughout their administrations for the ideals of conservation and for the expansion rather than curtailment of natural park and wilderness areas.

Still others, like Mrs. L. C. Adams in a postcard of 22 January 1954, simply called on the Secretary's official duty: "PLEASE protect Dinosaur National Monument. It is your duty to preserve such spots."

In the face of nearly 3,000 letters, Secretary McKay could respond only with a formula letter:

> My decision to approve the Echo Park Dam in the Dinosaur National Monument was not easy. Many factors had to be considered and a balanced judgment arrived at as to what was best in the national interest. As I am an ardent believer in our National Park System and intend to promote that system, the situation with which I was confronted was especially vexing. However, after careful study and review of alternative sites, it was my conclusion that the construction of the Echo Park Dam was necessary as a part of the comprehensive plan for the economic development of the Upper Colorado River Basin. So far as I am concerned, at least, I should like to make clear that my position in this case does not create any precedent for the construction of dams in any other National Park or Monument.[5]

McKay was playing the role of the consummate politician. He acknowledged the "vexing situation" and its complex of contested views and interests. His position, though, in the face of this first wave of letters remained steadfast. It would take a larger coordination of national groups to sway him. In combination with published books and letter-writing campaigns, the Sierra Club, the Wilderness Society, and the National Park Association, continually published updates and excoriations of the Department of the Interior, the President, and the Bureau of Reclamation in their house organs: *Sierra Club Bulletin, Living Wilderness,* and *National Parks Magazine.*

Starting in 1954 and continuing without abatement for a full year, in its *Bulletin* the Sierra Club published articles, editorials, and photographs of Dinosaur National Monument. The Sierra Club argued unrelentingly against the dam at Echo Park, discursively constructing the river as a natural aesthetic resource worth saving.

The Sierra Club called members to action with direct textual requests and by the presentation of images. The cover of the February 1954 issue, *Sierra Club Bulletin* carried an image of the Yampa River as it flowed through an area called Rainbow Recess in Dinosaur National Monument. Underneath was the headline in large font "Trouble in Dinosaur" and some short text describing the "primitive paradise unequalled anywhere . . . a unique gem of the National Park System . . . are needlessly threatened. You can prevent the destruction. Men of vision saved this place for us. Now it's turnabout." Underneath this text reads a large, underlined font "URGENT: Please read this issue now—and lend a hand."

The Wilderness Society, in a coordinated effort, sent the February 1954 issue of the *Sierra Club Bulletin* to their members with an additional message on the cover: "The Wilderness Society sends you this issue to stress the urgent need to act promptly." The lead article in this issue is entitled "Two Wasteful Dams—Or a Great National Park?" and argued forcefully against the need for them, contrasting this with the great inherent value of the place itself:

> Interior Secretary McKay is pressing for speedy authorization for Echo Park Dam, which would destroy the park value of the canyons of Dinosaur National Monument. . . . But are these two dams urgently needed? Are they needed at all? *Absolutely not.*

Highlighting the aesthetic value of the area, the article quoted the National Park Service in saying that "the effects upon 'irreplaceable . . . values of national significance would be deplorable'" (Sierra Club 1954: 3). The article continued by arguing that there were alternative sites, that the Secretary of Agriculture is currently worried about surpluses, and most importantly, that the "proposed Echo Park and Split Mountain dams would destroy the park value of Dinosaur; the unique would give way to the commonplace and would imperil the entire Park System" (*ibid.* 1954: 4). To the argument that the reservoirs would make the area more accessible to tourists, the author responded by pointing out that this would be true: "you can look at part of the setting [the highest 100 feet of exposed canyon]—after we've lost the priceless gem" (*ibid.* 1954: 4). The discourse used to defend Echo Park and Dinosaur National Monument continually reverted to a defense of the "priceless" aesthetic quality of the place. Nature, the Sierra Club argued, was irreplaceable, while the reservoirs had plausible alternatives.

Along with the traditional nature-photo trope showing deep canyons, striated cliffs, and serene flowing waters, the Wilderness Society's publication *Living Wilderness* added a photomontage that re-used techniques from the boosters of Boulder and Grand Coulee. Steamboat Rock, an immense monolith of sandstone, two thirds of which would be submerged by the reservoir was shown next to the Washington Monument (see Figure 6.3). The text accompanying the 1954 photomontage interprets the image as suggesting "a measure of the magnificence of the features preserved in the Dinosaur National Monument that are threatened by the proposed Echo Park Dam" (Wilderness Society 1954: 30).

Living Wilderness published many articles on the Dinosaur controversy, echoing similar discursive constructions of the Monument as a priceless scenic gem that should not be sacrificed for blind development. Adding to their photographic scenes of Dinosaur, *Living Wilderness* published a before-and-after collection of images from Hetch Hetchy, showing what a "developed" valley looks like.

Figure 6.3 Comparison of Monuments, *Sierra Club Bulletin* 1954.

In the text accompanying these photographs, the authors of *Living Wilderness* portrayed the Monument as being needlessly sacrificed: "This, as so much of the magnificent canyon country, will be inundated, awash, and lost forever, if Echo Park dam is built" (Wilderness Society 1954: 30). Much of this discourse was mirrored within the National Park Service itself in its fight to prevent the destruction of a National Monument.

The National Park Service, in an extraordinary conflict within the Department of Interior, fought strenuously against the Bureau of Reclamation plans for Dinosaur National Monument, dovetailing its arguments with the Sierra Club and the Wilderness Society. Though much of this conflict remained hidden from public view, some of it was apparent, and the results of the conflict can be seen in the subsequent restructuring of the National Park Service after the decision to drop the Echo Park and Split Mountain Dams.

Early in the process of developing the grand plan of the Bureau or Reclamation, the National Park Service appears not to have been concerned with the encroachment on the Monument that would occur due to the building of the two dams in Utah. In fact, a "memorandum of understanding" between the Bureau of Reclamation and the National Park Service, dated 4 November 1941, indicated that the "The Dinosaur National Monument region and its water control possibilities" is a most vital area for study.[6] Furthermore, "concerning the Dinosaur National Monument region, it seems not improbable that a policy similar to that already agreed to in principle for the Grand Canyon National Monument situation could be applied. Although legislation would be required in both cases to effect this policy, i.e., change the status of the areas from monument to recreational areas, the National Park Service does not believe such legislation would be difficult to secure." This change in status would allow development; a recreational area is a lesser category that does not limit usage the way that a National Park or Monument does.

Even through January of 1954, just before the Sierra Club issued its call to action, the Park Service was still interested in budgetary allocations from the Department of Interior so as to improve the section where the reservoirs would be located. In an internal memorandum, the Park Service estimated a cost of $21,000,000 needed to improve the area:

> Recreational opportunities and provision for their enjoyment on the reservoirs and adjacent lands will include picnicking, boating, fishing, swimming, hiking, riding, camping and sightseeing. These will require roads and trails to points of access on the reservoirs and to overlooks and view points. There will also be need for eating and overnight facilities, including a lodge and campgrounds.[7]

The Park Service was interested, no doubt, in making the best of a situation. At this point, the Secretary of Interior and the President were both set on moving forward with the Bureau's plans. In the face of this apparent juggernaut, the Park Service could at least capitalize on these plans by making the area accessible and developing it for maximum tourism.

Between 1949 and 1954, however, factions within the Park Service became more and more concerned about the precedent set at Dinosaur. Other Reclamation projects were being designed in or near National Parks or Monuments in Kings Canyon (California), Glacier Park (Montana), and the Grand Canyon (Arizona). Some Park Service officials feared a continuing loss of power vis-à-vis the Bureau of Reclamation. Reclamation already had a budget of more than five times the Park Service, and some Park officials worried that increasing their budget to develop recreational opportunities would not be worth the trade off of the precedent of inundating part of the Monument (Stratton and Sorotkin 1955).

As early as 1949 the conflict within the Park Service was apparent in some memorandums circulated inside the Service and even forwarded to the Secretary of Interior. One such memo stated outright that the Monument's "preservation in its natural state represents its highest use" (quoted in Stratton and Sorotkin 1955: 70). The Bureau and the Park Service continued attempts to balance their conflicting interests, and contradictory memos such as the 1949 and 1954 examples above must be seen in this light: they represent negotiations between several interest groups within the Department of Interior.

Just how much rancor existed between the Bureau and the Service is apparent in the accusations that the Park Service was behind the publication of Bernard De Voto's July 1950 article in the *Saturday Evening Post.* Michael Straus, then Commissioner of Reclamation, wrote to Newton Drury, Director of the Park Service, asking:

> Reclamation would be pleased to receive from the Park Service copies of all correspondence of the Service relating to and preceding the recent *Saturday Evening Post* contributions by Mr. De Voto that might throw light upon the source of certain of his statements as well as the photographs used so profusely in illustrating his assertions which were thoughtfully credited by the publication to the National Park Service.

These internal confrontations and accusations destabilized the Department of Interior and to some extent allowed Park Service personnel to coordinate with outside groups such as the Sierra Club and the Wilderness Society. For example, Park Service Director Drury wrote to a conservationist organization, regarding the Park Service's correspondence with the Bureau of Reclamation. This continued information exchange between the two agencies would, Drury hoped, "enable us to alert the conservationists of the Nation and more effectively with respect to remaining threats to national park areas from dam building" (quoted in Stratton and Sorotkin 1955: 76).

In 1953, just before the public controversy really exploded, some members of the Park Service also took a long-term view of this controversy. Commenting in a private correspondence that the conflict could actually improve the state of the conservation movement in the United States, one official wrote (quoted in Stratton and Sorotkin 1955: 75):

> I'm beginning to think the dam controversy will prove a good thing in the long run.

The recipient replied:

> I believe it has done more to bring the various conservation groups together than anything I can remember.

This assessment of the national situation was indeed accurate, perhaps even more so than the writers imagined. For in the next two years the public outcry took the shape of an oppositional discourse, in part defined by the National Park Service's defense of its preserved lands.

Conrad Wirth was promoted to Director of the National Park Service in 1952, and he proved an able manager of this conflict. After the controversy had subsided, in late 1955, Wirth started to outline his vision for the Park Service, a vision that would eventually come to be called "Mission 66," named for the goals that were to be met by 1966, the golden anniversary of the Park Service. The primary goals of Mission 66 were to bring the Park Service to the same funding level of other government agencies, such as the Bureau of Reclamation and the Army Corps of Engineers. Such agencies had long been able to secure multiple year budget allocations for long term projects (Wirth 1980: 239).

These longer term projects centered on improvements regarding "protection, public use, interpretation, development, staffing, legislation, financing, and all other phases of park operation that is in harmony with the obligations of the National Park Service under the Act of 1916" (internal memorandum quoted in Wirth 1980: 242). The first of the listed goals in the memo that Wirth wrote to the Mission 66 steering committee was protection of the National Park System. Given the timing of this program and the centrality of this goal, it is clearly a response to the proposed intrusion into Dinosaur National Monument.

The National Park Service had a public sphere group that advocated for protection and expansion of the Park System: the National Parks Association. In their quarterly periodical *National Parks Magazine,* rangers and officials of the National Park Service joined citizens and activists in writing about the "incomparable" loss that the Park System would suffer from the dam project, making them "useless for monument purposes" (National Parks Association 1954: 3).

The *National Parks Magazine* used the same environmental discourse as the Sierra Club and the Wilderness Society, with the additional legitimacy provided by the liberal inclusion of national and state park officials' statements, as well as pro-development voices, such as the Manager of Winter Park, Colorado, Stephen J. Bradley. Bradley wrote of his visit to Dinosaur National Monument in the April 1954 issue of *National Parks Magazine,* that "we were in a scenic area, the like of which for sheer dramatic beauty—of color, form, movement and sound—I had never experienced anywhere, and I have visited one-third of our National Parks." The discursive construction of the river canyons as an aesthetic treasure worth preserving was thus propounded from many sides, public and private.

In addition to their monthly publications and organizing efforts, the Sierra Club also internally distributed several policy manuals, guides for political action, and a "Public Relations Primer," with "how-to" procedures for contacting the press, making speeches, etc (Sierra Club 1957).

One section of this primer, entitled "misconceptions frequently encountered" lists a specific suggestion for responding to challenges such as "'The Sierra Club opposes progress; it is always opposing dams and roads.'" Readers were directed to respond with "The Sierra Club does not blindly oppose progress, it opposes blind progress. It opposes dams when it is proposed to build them in, or where they will affect, dedicated scenic wilderness and wildlife areas, especially when alternatives exist" (Sierra Club 1957: 9). Such clear training of its members helped the Sierra Club effectively oppose the dams at Echo Park and Split Mountain.

Unity of message, along with the many variants offered by members and affiliates in their letters to government officials, helped convince Secretary of the Interior Douglas McKay to drop the dams that would flood portions of Dinosaur National Monument. In late November of 1955 McKay announced his decision. Many letters of thanks followed.

Ralph Starr, a Sierra Club member, wrote to McKay on 30 November 1955, thanking him for his decision:

> The need for these magnificent sanctuaries has not diminished, their surroundings affect us all as visitors and their very presence assures us that we are a great nation sensitive to the beauties of the wilderness. The experience of the National Parks, and especially the grand ones of the West, is a pleasure that should be denied none of our citizens.

Such letters as Starr's were further reminders to McKay—as if after thousands of letters and acquiescing to public demands he needed more. Many letters made the same points about nature as magnificent, as unique, and as worthy of preservation in and of itself. Wilderness, the authors expressed, was not to be defiled in the name of progress.

Regarding the victory, David Brower noted in his diaries[8] on 1 December 1955:

> Recent events prove that people really believe in preserving portions of America's original beauty—and that the strength of their belief shapes policy. . . . But we need to rejoice with restraint, and need further reassurance.

The victory, for David Brower, was sweet. It certainly confirmed that Americans could exercise some form of direct democracy, and that enough of them believed strongly in preservation to sway the government. In short, he was witnessing the birth of a powerful new discourse—one that would electrify the environmental movements in the United States. It would, however, turn out to be a hollow victory for the Sierra Club and environmental organizations in general, and even a personal defeat of great significance to Brower. In order to remove the dams at Dinosaur, the Sierra Club was essentially locked into not protesting the great storage dam at Glen Canyon. Though the oppositional discourse at Echo Park focused on keeping the dams out of Dinosaur to preserve the wilderness area, the victory was won in part by showing how other aspects of the upper Colorado Basin development plan could substitute for the Echo Park and Split Mountain projects. In a letter to Secretary McKay on 20 May 1955, David Brower asked about increasing Glen Canyon Dam by 35 feet, pointing out that this could offset the loss of Echo Park and Split Mountain:

> Would it be physically possible to substitute for some of this storage? . . . by adding 35 feet to the present 700-foot height planned for the Glen Canyon Dam.

Through this strategy, Brower and the Sierra Club effectively shut themselves out of protesting Glen Canyon Dam. Many accounts at the time describe this as an agreement or a trade off, but there is little evidence of any formal pact. Instead, the Sierra Club had made a political mistake in granting the legitimacy of the Glen Canyon site and the upper Colorado project as a whole by suggesting raising the Glen Canyon Dam's crest height to offset the loss of Echo Park and Split Mountain. As Luna Leopold commented to Stegner and Brower: "if the Sierra Club gets into the problem of suggesting alternatives for Echo Park and Split Mountain Dam you are going to let yourself wide open" (Thomas 2000: 174). By granting this legitimacy the Sierra Club could hardly fight Glen Canyon Dam.

THE GLEN CANYON COMPROMISE: THE PLACE NO ONE KNEW

The victory at Echo Park was based on Brower's own presentation to Congress, where he made the explicit comparison to Hetch Hetchy, and told the Representatives not to make the same mistake twice. He also went on to show how Bureau engineers had failed in their math. Brower pointed out that the Glen Canyon Dam could be raised in height to increase capacity and thereby make up the loss of storage at Echo Park. The Bureau could thus back down on Echo Park while saving its upper Colorado development plan. It was a Faustian bargain for the Sierra Club though, as Brower soon understood, for Glen Canyon was an astoundingly beautiful place that few people knew about.

Glen Canyon was named by John Wesley Powell, the one-armed Civil War Army Captain who was the first white man, and perhaps the first human to raft down the Colorado, in 1869. In the area that would later mark the border between Utah and Arizona, Powell's exploration team drifted slowly between canyon walls that were up to 2000 feet deep, supporting a riparian ecosystem that was cool, green, and lush. As Lewis R. Freeman described it on his surveying journey in 1922:

> Glen Canyon is the Grand Canyon on a slightly reduced scale; but what it lacks in sheer magnitude it makes up in the added charms of its gentler natural beauties. Although its walls are neither less sheer nor less lofty than the average run of those in the Inner Gorge of the greater chasm, the less torrential current of the river—due to slighter declivity—in Glen Canyon has permitted the formation of more frequent and more fertile stretches of wooded bank and bench. One is never out of sight of trees, nor often of flowers . . . But the crowning glory of Glen Canyon is in the bowerlike amphitheaters of verdure that are responsible for its name . . . they form an almost unbroken chain of hanging gardens through the 150 miles from the mouth of Fremont River to the head of Marble Canyon.

This stunning beauty was so remote from Anglo society—there were several rough roads to it in the Navajo Reservation to the south, but none leading to the river from the north—that it was virtually unknown even halfway into the twentieth century.

It was in this "place that no one knew" that Glen Canyon Dam was built, begun in 1956 and finished in 1964. The Sierra Club mourned the loss with several publications, most notably the coffee-table book of Eliot Porter photographs published by the organization and edited by David Brower, *The Place No One Knew* (Porter 1963). In the Foreword to the book, Brower helped to entrench the environmentalist discourse that constructed the river and its immediate environs as a remarkable natural aesthetic resource:

> Glen Canyon died in 1963 and I was partly responsible for its needless death. So were you. Neither you nor I, nor anyone else, knew it well enough to insist that at all costs it should endure. When we began to find out it was too late (*ibid.* 1963: 7).

The building of a dam is equated with river death, and Brower admits culpability. After Secretary McKay's decision was made to remove Echo Park from development in favor of increasing Glen Canyon Dam's height, Brower went on several trips down that stretch of the Colorado, and described the area as some of the most magnificent scenery he had ever seen. Porter's camera recorded the beauty for other Americans to see.

Eliot Porter wrote much of the text that accompanied his photographs. When combined with the images, his words repudiate the discursive construction of the river as a "menace" by the Bureau. The river is characterized as "serene" and "overwhelming:"

> The eye is numbed by the vastness and magnificence, and passes over the fine details, ignoring them in a defense against surfeit. The big features, the massive walls and towers, the shimmering vistas, the enveloping light, are all hypnotizing, shutting out awareness of the particular (*ibid.*1963).

The superlatives in his text are easily matched by the photographs, printed in stunning clarity and color (see Figure 6.4 reproduced here in black in white). Porter's viewer described in the text lingers on the general features for only the first moments, and is soon caught up in the finer details that were originally overlooked. The photographer turns from the wide-angle to the close-up and an intimacy of the canyon becomes apparent:

> Then you see for the first time the velvety lawns of young tamarisks sprouting on the wet sandbars just vacated by the retreating flood, or notice how the swirling surface of the green, opaque river converts light reflected from the rocks and trees and sky into a moiré of interlacing lines and coils of color, or observe the festooned, evocative designs etched into the walls by water and lichens (*ibid.* 1963).

Porter continually moves between the macro and the micro in his text and his photographs. He records what this "place that no one knew" was like before it was inundated. The larger picture that emerges from the collection of images and text is that of a tremendous aesthetic asset that has been lost. The work argued forcefully, if indirectly, against the discursive construction of the river as a menace, a tyrant, or an agent of chaos.

Another author who experienced the river before the dam was put into place was Edward Abbey. The book that brought fame to Abbey was *Desert Solitaire* (1968), which contains a chapter called "Down the River." This piece describes Abbey's raft trip through Glen Canyon all the way to the site of the dam, just as the foundation was being poured. It opens with Abbey's characteristic acerbity:

> The beavers had to go and build another goddamned dam on the Colorado. Not satisfied with the enormous silt trap and evaporation tank called Lake Mead (back of Boulder Dam) they have created another even bigger, even more destructive, in Glen Canyon (*ibid.* 1968: 173).

Figure 6.4 Coffee Table Epitaph: *The Place No One Knew.* Eliot Porter, *Burned Trees*, Glen Canyon, UT, 1961 color transparency. Courtesy the Amon Carter Museum.

Abbey's chapter continues on, using corrosive sarcasm, to belittle the Bureau of Reclamation and the federal government in general. In the midst of this rant, Abbey finds time to describe the scenery in the canyon, which will be submerged under 400 feet of water in a matter of months: "white sands," "green willows," "a sculptured landscape."

Later in the chapter, as Abbey and his rafting mate begin to come to grips with the fate of this canyon, Abbey dreams of the opening ceremony:

> Some unknown hero with a rucksack full of dynamite strapped to his back will descend to the bowels of the dam; there he will hide his high

> explosives where they'll do the most good, attach blasting caps to the lot and with angelic ingenuity link the caps to the official dam wiring system in such a way that when the time comes for the grand opening ceremony, when the President and the Secretary of the Interior and the governors of the Four Corners states are in full regalia assembled, the button which the President pushes will ignite the loveliest explosion ever seen by man, reducing the great dam to a heap of rubble in the path of the river. The splendid new rapids thus created we will name Floyd E. Dominy Falls in honor of the chief of the Reclamation Bureau (*ibid.* 1968: 188).

Abbey's disdain for the governmental agencies and the Bureau in particular, is immense. In his later books, *The Monkey Wrench Gang* (1975) and *Hayduke Lives!* (1990), Abbey's characters famously set about achieving his own lifelong dream of blowing up Glen Canyon Dam. Although Abbey did not contribute anything directly to the debate around Glen Canyon Dam, as one of the final visitors to the place no one knew, he bore witness to the aestheticized nature that Brower and Porter mourned.

The Navajo Indians also bore witness to the damming of Glen Canyon. The Navajos were not politically well organized in the mid-1950s. They had recently suffered a great deal of hardship due to Bureau of Indian Affairs herd reduction programs in the late 1930s (White 1983: 313). (Interestingly, the herd reductions were intended to decrease erosion of their lands, erosion that flowed directly into Lake Mead and silted up behind Boulder Dam. The herd reductions were imposed precisely because Department of Interior officials feared a loss of efficiency at Boulder.) This social disarray led to an official, but essentially meaningless, approval of Glen Canyon Dam by the Navajo Tribal Council. The Council, a group that was formed at the behest of the Bureau of Indian Affairs to handle the sale and leasing of rich oil lands on the reservation in 1927, had very little legitimate governing power over the diverse and widespread peoples of the reservation. They did, however, pass motions approving proposals from the Bureau of Indian Affairs (such as herd reductions) and the Bureau of Reclamation (Glen Canyon Dam). Raymond Nakai, the Chairman of the Tribal Council at the time that the dam was built, said "A conservationist is one who is content to stand still forever. Major Powell would have approved of this lake. May it ever be brimmin' full" (McPhee 1971: 196).

Nakai's comments notwithstanding, many Navajos did not approve of the dam, though little of this contestation made it into the historical record. There was significant disagreement about the benefits that would flow to the Navajo. There were many who, gesturing to history, asked what the Navajo had ever seen of other promises made by the government. Others believed the Bureau of Reclamation promises of hydropower and irrigation water (John 2000). In the end, the dam was built, based in part on the approval of the Tribal Council. The Navajo have, forty years later, not received any benefits from the dam aside from increased tourism in the area.

Though Abbey and the Navajos did not directly attack the Bureau's discourse, Brower, in the Foreword to *The Place No One Knew,* did argue against the discursive construction of the Colorado as an economic resource:

> Good men, who have plans for the Colorado River whereby "a natural menace becomes a natural resource," would argue tirelessly that the Colorado must be controlled, that its energy should be tapped and sold to finance agricultural development in the arid West. But our point here is that for all their good intentions these men had too insular a notion of what man's relation to his environment should be, and it is tragic that their insularity was heeded. The natural Colorado—what is left of it—is a miracle, not a menace. The menace is more likely the notion that growth and progress are the same, and that the gross national product is the measure of the good life (*ibid.* 1963: 7–8).

Brower met the Bureau's arguments head-on. The river was not, Brower contended, a menace. The menace instead was that constellation of forces that push for what Brower called "blind progress"—those that rate value only by economic measures. Brower clearly showed his bitterness in this work: his comments regarding "what is left of" the river, the selling of energy to finance agriculture, the argument for storage that Brower asserts is "absolutely not needed in this century, almost certainly not needed in the next" (*ibid.* 1963: 7). Brower here was resigned to the loss, yet vowed never to let something like this happen again:

> The Sierra Club has no better purpose than to try to let people know in time. In Glen Canyon we failed. There could hardly be a costlier peacetime mistake. With support from people who care, we hope in the years to come to deter similar ravages of blind progress (*ibid.* 1963: 9).

In the years to come, Brower and the Sierra Club would indeed have occasion to wage battle against the Bureau's developmentalist mindset. They fought hard to lower the storage level of Glen Canyon so as to avoid the inundation of Rainbow Bridge National Monument. The original height of Glen Canyon Dam was to be 580 feet, but with the compromise it was raised to 730 feet. When it was found that this would bring the water level to the base of Rainbow Bridge, one of the natural wonders of the world, the Sierra

Club lobbied for protection. Many schemes were put forward, including a check dam below Rainbow Bridge to keep the reservoir water out. Eventually the Bureau accepted a proposal to keep the water level of the reservoir at 3700 feet above sea level, and to build the dam to just 710 feet high.

The Sierra Club's loss at Glen Canyon was a high price to pay, but the failure in many ways galvanized the new environmentalist discourse in the United States. Since the building of Glen Canyon Dam no more high dams have been built in the U.S. The Sierra Club was successful in stopping several more dams on the Colorado, including two in Grand Canyon. Brower took out several ads in the New York *Times,* the first overtly political advertisement in that periodical. The IRS stripped the Sierra Club of its tax-exempt status for these actions, but the Club gained so many new members that the losses were largely offset.

The Club has continued to grow in strength and purpose over the nearly four decades since the "death of Glen Canyon." In 1993 the Club attempted to avenge its loss of Glen Canyon by helping to write legislation to breach the dam and restore Glen Canyon. The legislation remains stalled, but the discourse continues to be a powerful force in America.

COOPTING THE ENVIRONMENTALIST DISCOURSE: THE JEWEL OF THE COLORADO

In 1962 the U.S. Bureau of Reclamation produced a film called "Canyon Conquest," which tells the story of the construction of Glen Canyon Dam. Using all of the tropes of the imperial modernist discourse, "Canyon Conquest" strikes the modern viewer as an anachronism. With its deep, authoritative male voiceover and its blaring march music in the background, the film recounts the conquest and domination of the Colorado River.

The film opens with a close up of a Navajo man gazing into the distance, with a soundtrack in the background that sounds strikingly like the overture to the Rogers and Hammerstein musical *Oklahoma!* The camera pans out to show grazing sheep, and the voiceover resoundingly says:

> Lines deep-etched on the face of Alvin Chinniginny speak of the desert sun, of the sand-flecked wind, of the grinding fight for survival. For Chinniginny and his fellow tribesman, life has been harsh indeed, linked always to the parched wind-blown soil, linked to the small flocks of sheep scarcely large enough to sustain life: A way of life resembling that of Biblical times.

This noble-savage caricature of Chinniginny is a foil that is used throughout the film. Chinniginny is taken on a tour of the dam site and shown the wonders of modern technology, "just twelve miles from his hogan of grass and mud." This imperialist and high modernist discourse not only constructs the area as harsh, forbidding, and arid, but it also obscures important political and social realities. The sheep that Chinniginny is shown tending in the film are a small flock indeed. This is not, however, due to the climate, the land, or to Chinniginny's incompetence as a herder, as the film suggests. Instead, it is a direct result of the 1933 herd reductions that were forced on the Navajo Nation by the Bureau of Indian Affairs (BIA). The trope of the noble savage living on the edge of subsistence is used as a reason to bring civilization to the region:

> For Chinniginny and his fellow tribesmen this was to be a different day, different in a way that he could scarcely imagine. For this was the day that men in a jeep bounced across the mesa to the canyon called Glen. Men who were to bring the twentieth century to Chinniginny.

The twentieth century is, then, to finally come to the Navajo and the Southwest. If asked, most Navajo would likely have responded that the last time the twentieth century was brought to them (in the form of herd reductions) it caused deep misery, famine, and death. Glen Canyon Dam and the Anglo's twentieth century would have been suspect to many Navajo.

The film continues, using Chinniginny's tour of the jobsite to show the progress that the dam will bring and the miracles of modern science and technology that the construction process exhibits. The film uses the standard arguments of the imperialist high modernist discourse, calling the river "untamed" and "rich in beauty, rich in potential." Chinniginny tours the jobsite, seeing this potential being harnessed and tamed. The benefits of the dam are described to Chinniginny:

> Colorado River water could mean energy; properly controlled and utilized could mean power, pulsing, flashing energy for cities virtually unborn; could mean irrigation for the thirsty crops of upstream valleys; life for the swelling populations as yet unborn.

These standard arguments of the imperial modernist discourse are given hyperbolic expression in the film "Canyon Conquest." The film ends by showing Chinniginny riding his horse-drawn wagon home to his hogan, while the voiceover says, "A way of life passes, a new way is born." The meaning is clear, though not stated outright: the Native American way of life is outdated, and the technological progress of Anglo-America is transcendent. This film, racist as it is, is a product of its time. The producers clearly felt that history was progressing in an obvious direction and that they were on the right side of it. What was left behind, however, was the discourse that the film used, not the way of life of the Indians. The Bureau of Reclamation was operating in an automatic manner. After the failure of such discursive techniques to gain support for the Dam, the Bureau had to try something new.

In response to the publication of such funereal works as *The Place No One Knew* in 1963, just after "Canyon Conquest," the Bureau of Reclamation had to engage in another discursive construction of the river—now reservoir—as a benefit to all Americans. This reconstruction was accomplished through the creation of a new discourse based on the improvement of nature. This discourse highlighted the benefits of the dam and the reservoir, showing how the increased availability of the area for the purposes of recreation was an unmitigated improvement upon what had existed prior to the dam. This discourse was a direct response to the oppositional discourse at Glen Canyon, which argued for the aesthetic value that inheres in nature. The *improved nature* discourse argued forcefully that the results of human labor were better than what was there before. The Bureau of Reclamation used several different techniques to make this argument, including glossy pamphlets, motion pictures, and promotion through hiring a famous painter to produce representations of the dam.

The 1965 production of a color film called "Lake Powell: Jewel of the Colorado" coincided with the publication of a book by the same name (see Figure 6.5).

Figures 6.5 and 6.6 Jewel of the Colorado, U.S. Bureau of Reclamation, 1965

In the movie a helicopter tours the Reservoir, showing sweeping vistas of blue water with red cliffs in the background. The book reproduces film images showing families camping, water skiing, and recreating (see Figures 6.6 and 6.7).

The work portrayed the reservoir as an unmitigated benefit. Much of the text accompanying the images was in poetry format, emphasizing the peaceful, recreational, and religious motifs of the work. Several pages of narrative accompany the photographs, written by Commissioner of Reclamation Floyd Dominy, Secretary of the Interior Stewart Udall, and Bureau staff writers. Each writer serves different purposes. Dominy highlights the recreational aspects, calling on readers to explore, hike, water ski, swim, and, most importantly, find peace of mind and religious fulfillment:

> If you're tired in mind and soul, in need of restful serenity, I don't know of a better place. If you want to be alone, you can be alone. . . . You have a front row seat in an amphitheater of infinity. . . . There is peace. And a oneness with the world and God (Bureau of Reclamation 1965: 16).

Dominy's emphasis on recreation was joined with Udall's discursive construction of the area as one of outstanding natural beauty—obscuring the fact that there was another world of beauty under 400 feet of water. "Once in a blue moon we come upon almost unbelievable beauty. Such was my reaction at my first sight of Lake Powell and its setting of incomparable grandeur" (*ibid.* 1965: i).

Other Bureau writers, using the standard arguments of imperial modernist discourse, took up the many benefits that Lake Powell was to provide for Americans: water for irrigation, drinking water, and hydropower. One author makes a strong case for altering nature:

> There is a natural order in our universe. God created both Man and Nature. And Man serves God. But Nature serves Man. Man cannot improve upon Nature. But—as he has since before the dawn of history—Man must continue to adapt Nature to his needs. Still, that process of adapting must preserve—in balance—the whole natural heritage that is his. The Colorado River and its basin are a great and abundant treasure house of natural resources and natural wonders. Let us husband the one wisely. Let us enjoy the other fully (*ibid.* 1965: 28).

This passage shows the effects of the powerful environmentalist discourse on those who opposed its goals. The Bureau of Reclamation shifted its own discourse in the face of such effective opposition. It now made a case for altering nature based on economic, religious, and historical legitimacy. The rise of the discourse of environmentalism would not be fully complete without a demonstrated response from the Establishment.

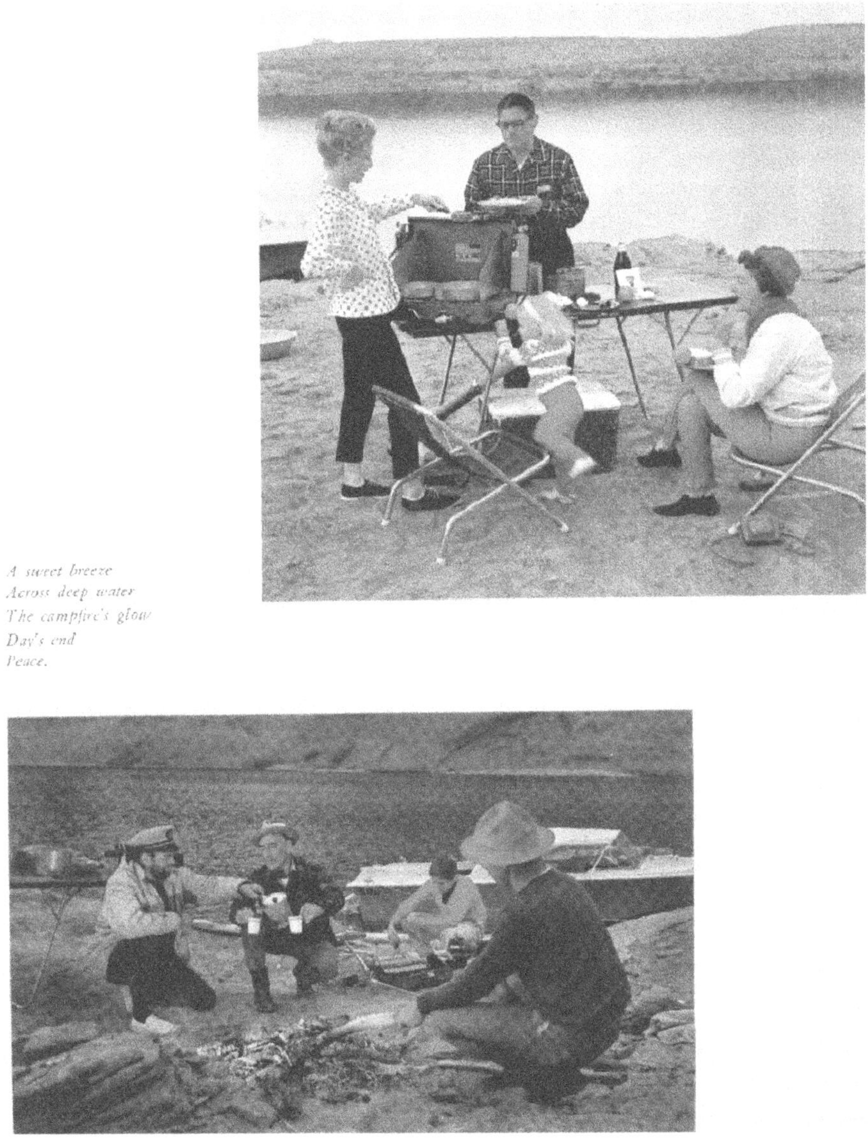

Figure 6.7 Re-Creating on the Colorado, U.S. Bureau of Reclamation, 1965

To support these arguments for improving nature, the Bureau hired Norman Rockwell to paint the dam (Figure 6.8). Rockwell visited the dam site and reportedly said, "That is a lovely thing, but I don't paint things, I paint people." Bureau personnel recruited a Navajo family to pose in the foreground of the dam, and Rockwell could paint a human element into his work.

Figure 6.8 "Glen Canyon Dam" by Norman Rockwell, Oil on Canvas.

Rockwell's painting used many of the standard tropes of his other works, showing the splendor of the dam along with a symbol of American freedom (the bald eagle) and the Navajo family, which stands as the focus of the painting, but not its main subject. The dam stands behind and to the left of the family, who gaze at what the viewer must imagine is the symbol of their new freedom in front of the vast expanse of nothingness beyond. The dam, of course, did not free the Navajo from their complex dependency on the larger American economy and society (White 1983). Rockwell's painting, however, did help to legitimize Glen Canyon Dam as an important symbol of American life. Rockwell, of course, represented the wholesome middle-America in oil on canvas: His covers for the Saturday Evening Post defined what was good and right about post World War II America to many of the newly affluent. Just as they may have seen "Going and Coming" (1947) as an endorsement of the consumption of aestheticized nature in the 1940s, in the middle 1960s America must have seen the painting of Glen Canyon Dam as a similar endorsement of the developmentalist ideas of the Bureau of Reclamation, and an attempt on the Bureau's part to gain legitimacy.

The discourse on improving nature is thus a response to the aestheticization of nature by the opposition groups in the late 1950s. By reviving John Muir's discourse regarding the inherent value of nature, the environmentalist discourse found a powerful base upon which to contest the "forces of blind progress" that constructed nature wholly in terms of its economic potential. The powers that be, in this case primarily the Bureau of Reclamation, responded by valuing that which is built by society. The Bureau attempted to show the greater value in recreational facilities, in improved access, and in economic benefits that accrue after a dam is constructed. This discourse of improved nature was an attempt to incorporate the oppositional discourse of environmentalism and thereby derogate the oppositional discourse. If the product of the human domination of nature is itself as useful and beautiful as wilderness, then the environmentalist discourse is not legitimate.

CONCLUSION: EFFECTIVE OPPOSITION AND THE COMPLEXITY OF HEGEMONY

Through the struggle to save Dinosaur National Monument, and in the mourning of Glen Canyon, a new oppositional discourse became enshrined. This oppositional discourse, which I have labeled environmentalist, constructed nature as a priceless treasure that needed to be protected from blind progress. Drawing on the lessons learned from Muir's failed Hetch Hetchy battle, activists successfully fought against the intrusion into the National Park System by the Bureau of Reclamation. They were able to discursively reconstruct nature—in the form of the Colorado River—as an entity that had value in and of itself, not simply something that could be economically beneficial to society.

In a dramatic shift from the lack of oppositional discourses regarding the proposal of Boulder and Grand Coulee, a powerful new discourse was born in the struggle over the proposal of Echo Park and the building of Glen Canyon Dam. This discourse highlighted an inherent value that existed in nature and wilderness, a value that was worthy of preservation over exploitation.

This shift in the valuation of nature points to a larger transformation of the relationship between nature and society. This transformation is certainly something that is in process, unevenly completed, and perhaps will never be as hegemonic as imperial modernism was. This new view of the nature-society relationship, a view based on preservation and inherent natural value, has destabilized the hegemonic imperial modernist ideology. In offering an effectual foundation for oppositional discourse, the environmentalist ideology has stopped the damming of the rivers of the West, and forced society to re-examine its relationship to growth, natural resources, and state building.

Some commentators have identified the Echo Park-Glen Canyon episode as a significant moment in the development of environmentalism in the late twentieth century. Though the discourses used were not new, they were mobilized on a massive level and in an effective way for the first time. Gottlieb (1993: 41) notes the historic significance of this battle over wilderness in his book on the origins of the modern environmental movement, and Mark Harvey (1994) picks out Echo Park as a "symbol of wilderness" that was "a great test to the sanctity of the park system" (Harvey 1994: xiv). Though the use of the singular "movement" by Gottlieb should be questioned, the importance of Echo Park and Glen Canyon for the set of environmental movements that have blossomed in the last forty years is clear.

The imperialistic modernist discourse about the domination of nature and the building of a state infrastructure in the West was transcendent from at least the early 1920s through the middle 1950s. This hegemonic discourse brought together aspects of state building and the domination of nature (control of nature, control of the population, and the boosting of economic development) into a monolithic discourse that defined the relationship between nature and society. With the rise of an oppositional discourse at Echo Park and Glen Canyon, this hegemony was first successfully contested. Hegemony, of course, thrives on the absorption of opposition. This cooptive ability can be seen in the support of Grand Coulee Dam by Woody Guthrie, and the compromise between the Sierra Club and the Bureau in regards to Glen Canyon.

The ideology of imperial modernism, however, cannot easily survive a decentering such as it experienced with the introduction of the environmentalist discourse. Though the powers-that-be crafted a cooptive discourse of the improvement upon nature, it is questionable how well this discourse will take hold. Perhaps we shall see modernism absorb this opposition as it has others before it. Perhaps imperial modernism shall not remain hegemonic. This future is open.

And so it must be noted that the environmentalists should not rest easy. The oppositional discourse of preserving aesthetic nature is a powerful one. It has brought the environmental movement in the U.S. to the bargaining table of politics, and important victories have been won. Yet the Bureau of Reclamation and other political agencies have reacted to this discourse in immediate and substantial ways. The cooptation of the environmental discourse may lead to the undoing of the environmental movement in the future. If the "agents of blind progress," as the Sierra Club once called the Bureau of Reclamation, coopt the rhetoric of aesthetic value in nature, and the environmentalists are not careful, they could find their own discourse used against them in the future.

Chapter Seven

DamNation: Controlling the Waters, Civilizing the Wilderness

> We can say this: That what we call "development" is not a uni-directional process, especially in a semi-arid country. To develop this land we have used engines that we could not control, and have started actions and reactions far different from those intended. Some of these are proving beneficial; most of them harmful. This land is too complex for the simple processes of "the mass-mind" armed with modern tools. To live in real harmony with such a country seems to require either a degree of public regulation we will not tolerate, or a degree of private enlightenment we do not possess.
>
> But of course we must continue to live with it according to our lights. Two things hold promise of improving those lights. One is to apply science to land-use. The other is to cultivate a love of country a little less spangled with stars, and a little more imbued with that respect for mother-earth—the lack of which is, to me, the outstanding attribute of the machine age.
>
> Aldo Leopold, 1933, "The Virgin Southwest"

> Standing on the edge
> of the Hoover Dam,
> I'm on the centerline
> right between two states of mind.
>
> Bob Mould, 1992, "Hoover Dam"

Elwood Mead, Commissioner of the U.S. Bureau of Reclamation from 1924–36, once quipped that even if all the rain that fell on the mountains of the West were captured, it still would not slake the thirst of the farmers west of the 98th meridian (Mead 1903/1972: 5). In a half-century attempt to capture this water, the United States developed almost all of the rivers in the 11 western states. Using discourses of imperialistic high modernism to justify their program of damming the western half of the nation, the Bureau created an infrastructure to stimulate the economy and society of the West. As Aldo Leopold noted in the epigram above, this engine of progress was, for fifty years, a Frankenstein's monster. Some of the intended results were certainly beneficial; other unintentional outcomes were extremely harmful. The engine of progress included a discourse of imperialistic high modernism that justified the domination of nature on a massive scale in the American West. The forces of "blind progress," as David Brower called them, were finally stopped at the battles over Echo Park and Glen Canyon in the late 1950s. For many observers, this partial shift in the alignment of powers from the state to civil society marked the beginning of a new environmental movement in the U.S.

This was not, however, the first time that the voices of civil society had been heard. Though it was the first time that effective opposition was mobilized, civil society groups had been central to the building of Grand Coulee Dam. Many of the ideas of the Progressive Era, such as a faith in science and technology to solve social and environmental problems, a belief in linear progress, and the correlation of efficiency with social good, were translated into materiality in the American West not through the heavy hand of the state, but instead through the vocal demands of civil society.

STATE BUILDING IN THE AMERICAN WEST: CULTURE, GEOGRAPHY, HISTORY

Geographic, political, and economic influences can be seen in the three cases of state building exhibited in this work. The vast expanses of land and the extreme geography of the West—especially its aridity—were important to the process of state building and help to differentiate it from cases in the eastern U.S. and in early modern Europe. To overcome the problems of aridity, the state had to construct an infrastructure that would provide irrigation and drinking water, produce hydroelectric power, and reduce flooding in the West. This form of state building relied upon a cultural and discursive construction of the relationship between nature and society based in domination. To build the dams, the state and local citizens had to justify domination of nature for human economic purposes. This occurred both through the use of state simplifications to order the landscape and through a discourse of imperialistic modernism that legitimated such orderings and the general domination of nature.

The discourses around Boulder and Grand Coulee precluded any discussion of not building the dams. The competition between discourses was instead over who would build the dam and what it would look like: Not only did the state-sponsored, imperialist modernist plan win approval in both cases, but those who initially opposed a state-sponsored plan eventually backed it in order to get their part of the benefits. Even those one might imagine would protest the dam supported it. The cases are similar in outcome, but different in process: At Boulder it was primarily the U.S. Bureau of Reclamation, Congress, and Imperial Valley farmers who championed the dam, while Grand Coulee saw a large array of civil society groups fighting to convince Congress to finance the dam and the Bureau to build it.

The imperial high modernist discourse was key to this process in both cases, especially as used by boosters such as U.S. Representative Scott Leavitt, Republican of Montana. Leavitt, in hearings regarding Grand Coulee Dam in 1932 (chapter five), compared the project in worth and importance to the building of the Panama Canal. He connected dam building with rhetoric about empire building, national defense, and the orderly growth of the population in the West (House of Representatives 1932):

> The orderly development of our population and resources on the west coast is of extreme importance; and the way to develop is through the development of those irrigated sections that become the center of all industry and livestock raising and so on. We have here, of course, a combination of power development and the development of a great area for tremendously aiding the growth of population on the coast. We will never be fully developed, from a national defense standpoint, until the west coast is as fully developed as the east coast.

Leavitt's statement to the House Committee on Irrigation and Reclamation is an archetypical example of the imperial high modernist discourse. Leavitt described the need for orderly development, population growth, progress, and imperialism.[1] This development was to be achieved through the rational application of science and technology to the landscape—by building a dam. The imperial high modernist discourse was used to great effect, culminating in the Bureau of Reclamation's grand plans to develop the entire Colorado River in its 1946 master plan (chapter six).

The Bureau's plan to completely subdue the Colorado exposed its folly and ended the "go-go years" of dam building in the West, as civil society groups such as the Sierra Club organized against the plans to construct a dam in Dinosaur National Monument. The oppositional environmental discourse that arose contested the imperial high modernist discourse by claiming inherent values of nature: aesthetic, spiritual, and comparable in worth to any economic development.

The three cases in this work also show the powerful role of civil society in shaping the process of state building in the American West. Boosters in north central Washington and southern California pushed their local projects, often pitching them in ways that highlighted the benefits to the nation at large. This pressure helped to keep the projects in front of administrators and bureaucrats, as well as elected officials, ultimately helping to secure funding from the federal government.

To the contrary, at Glen Canyon civil society groups nationwide fought against the Bureau of Reclamation, keeping it from building two dams in Dinosaur National Monument. The class basis of this effective opposition lies solidly in the burgeoning middle classes of the 1950s. Automobiles and highways brought many in the middle classes into contact with nature, allowing them to develop conservation sensibilities. Opposition groups tapped into these nascent sentiments, expressing and refining them into a political force.

The linkages and contests between civil society and the state in these cases are not reducible to a simple dichotomy or a hierarchy of influence. At times, civil society groups fought one another, the state, and elites over the process of state building. The state's attention to the projects waxed and waned (Grand Coulee), local elites did not always get their way (Boulder and Grand Coulee), and sometimes the elites and the state lost entirely (Glen Canyon). Hence neither a state-centered theory nor a class-based theory explains the case of state building in the American West. Instead, this work supports "polity-oriented" theories that span a middle ground (see, for example, Evans 1995; Gilbert and Howe 1991; Skocpol 1992).

Cultural influences go a long way in explaining the variations in state building across different regions and nations at different times. Though many of the same processes that influence state building in general, as identified by Skowronek (1982), were in play in the American West, additional cultural influences are also identifiable. The particular cultural constructions of the relationship between nature and society by local inhabitants had an important role in supporting or contesting the state-building process. The discourses of high modernism helped legitimate the domination of nature and enabled state building. As many scholars have shown—starting with Weber—culture is how legitimation happens. Culture, either through tradition, custom, or elective affinity, can be a powerful justification for domination (Weber 1946: 285). But culture is also the medium through which dissent is expressed, as can be seen in the sarcasm in Woody Guthrie's songs or the ballistic tone of Bernard DeVoto's essays.

This work has shown that discourses regarding nature and environment have an important role in the historical process as they are attached to practices and thus operate in material ways. The hegemonic discourse of imperial high modernism was based in dominant systems of power and thus superceded and/or absorbed opposition. The case studies have shown how the state, through the building of an infrastructure, enabled continued capital accumulation. This accumulation was, then, based on the domination of nature, the absorption of opposition, and the discourse of progress, efficiency, and rationality.

This work has presented a discursive analysis of the process of state building (state building so as to control population, territory, and enable the accumulation of capital) and has focused on the ideas and practices of the various groups of actors. State officials, local and regional elites, the press, and activist groups all contributed to the discourse about the building of a water infrastructure by the state in the American West. These multi-vocal and polyvalent contributions show, at different times, both a strong civil society and a strong state. The analysis of historical discourses shows that high modernism does not inevitably lead to the overextension of state action. In the case of Boulder Dam the state did act in ways that are close to a pure case of high modernism, but in the case of Grand Coulee Dam it was the civil society groups who pushed the state to act, while the final case of Glen Canyon Dam, activist groups kept the state from building a set of dams.

This work has also shown the importance of cultural constructions of nature—constructions of water as something to be put to use, and of a river as something to be dominated. Both state and civil society groups supported the material culture of expanding an economy, building state institutions, and engineering physical structures. These discourses expressed a particular understanding of the relationship between nature and society whereby nature existed solely for human use and the consequences of this domination was straightforward. While many historical actors understood this relationship as a simple and direct one, it was (and still is) complex and dialectical, often with negative unintended consequences.

A TREADMILL OF PRODUCTION ON SOUTHERN FRONTIERS

In chapter two, I suggested that one way to understand the building of water projects as part of a social and economic development process was to use Schnaiberg's (1980) concept of the treadmill of production. The development of the American West shows how a treadmill of production was mounted by the Federal government in alliance with individual states, private capital, workers, and local elites in order to ensure economic and social stability and to enable continuing growth. That this treadmill of production is based on a tight coalition of mutually interested groups that comprise powerful segments of society underscores Schnaiberg's original conclusion that, once mounted, the treadmill is difficult, if not impossible, to dismount.

The dams of the West were a model of state building where the government put up money and planning for massive infrastructural improvements that benefited private capital, cities, and farmers. This model has been exported to many developing nations as early as the 1950s, and continues to this day.

Chile's Biobio River and the network of dams that are under construction or have been proposed exemplify the transplanted technology and ideology better than most. The river flows from the peaks of the Andes to the Pacific in central Chile: this massive elevation drop is proposed to be harnessed to provide energy to be used locally and exported to regional markets, will provide recreation, and will offer drinking and irrigation water for the nation's farmers and cities. Six hydroelectric dams were proposed for the course of the river in the late 1970s under dictator Augusto Pinochet, and have been pursued under the proceeding neo-liberal regimes, with backing from the International Monetary Fund and World Bank. One dam was completed in 1996 and construction on the second was begun in 1999 (International Rivers Network 2003a).

Critics of the project include international groups and indigenous peoples from the upper reaches of the river who have been or will be displaced (LaFranchi 1998). Refusing to leave, and suing in Chile's courts based upon weak precedents, the Pehuenche people have been a lightening rod for intentional criticism of the projects. Joining the human rights observers are whitewater rafters from many nations who want to protect some of the world's most famous and challenging routes, as well as environmentalists who wish to save several unique species of fish that the dam would drive to extinction.

The dam presently under construction, at Ralco, will provide 1/5 of central Chile's energy needs, including feeding the voracious energy appetite of Santiago, which has doubled its energy use each recent decade. This is the central goal of the growth coalition surrounding the Biobio project, and one that will likely not easily be opposed. On 1 October 2003, the four main indigenous women opposing the dam through civil disobedience gave in to the pro-dam forces and accepted an offer of nearly $300,000 each and relocation to 77 hectares (190 acres) each of land grants (González 2003).

Similarly, India's Sardar Sarovar Dam on the Narmada River in the Madhya Pradesh state of central India is being constructed for energy production and to secure a stable source of irrigation and drinking water in a region where most ground water is available only immediately following a monsoon (Sardar Sarovar Narmada Nigam Ltd 2003). This dam is the centerpiece of a 30 dam proposal that covers the length of the Narmada River through three states.

Just after Independence, in 1948, Prime Minister Jawaharlal Nehru famously said to villagers who were to be displaced by a dam, "If you are to suffer, you should suffer in the interest of the country." India's passion for dam building has led to many dubious projects, but many successful ones as well. All the projects are done in the name of power, flood control, and especially drinking and irrigation water. India is perhaps the international poster-child for a state-led treadmill of production: some 3600 dams have been built in the space of 50 years.

Critics of the Sardar Sarovar Dam extend throughout India and to an international community, famously including the prize-winning author and activist Ahrundhati Roy. The central issue of contention is the resettlement of hundreds of thousands of people. The Sardar Sarovar Dam is, essentially, the straw that broke the camel's back—though it is quite large and quite destructive, it follows the construction of 3600 other large dams across India, each of which has displaced an average of 44,000 people, or about 50 million people since Independence (Roy 1999). Though dams have been protested in India over the last 20 years, none have mobilized the multiple criticisms based in ideas of social justice, environmental sustainability, economic efficiency, and cultural survival that the present groups have used, and no past protests have attracted attention through celebrity activists such as Roy (Guha 2000). That the high profile movement can stop the dam is unlikely (the dam is currently nearly 110 meters high, and is planned to reach 163 meters), though it may be successful in lowering its height and reducing its negative impacts.

The Three Gorges Dam on China's Yangtze River is shaping up to be the world's largest and most expensive public works project in history. With an initial estimated cost of $24 Billion the dam eclipses all known projects—and estimates of its eventual cost range up to $75 Billion. The dam was originally proposed as early as 1919, with debates reinvigorated in the 1940s and 1950s. The dam was officially approved for construction in 1980, with construction begun in 1994. Though at times there was intense conflict over both the feasibility and desirability of the dam, the government denies any conflict (China Online 2003; The Embassy of the People's Republic of China in the United States of America 2003). The dam is, of course, being built by a Socialist government (albeit one that is increasingly opening to world markets), which makes it an interesting counterpoint to the American, neo-liberal, and post-colonial examples above.

The Chinese government suggests multiple benefits from this development: hydropower production, improved navigation, and flood control. The new navigation route could reduce transportation costs by up to 30% and periodic flooding (which has plagued the area for centuries) would be controlled. Significantly for the global environment, the massive hydropower generation will reduce China's reliance on coal burning power plants.

Critics of the dam, both internal to China and internationally have taken issue with just about every one of these claims. International critics claim that siltation will render new ports unusable in a matter of decades (International Rivers Network 2003b), and that there is no market for the electricity that is already being produced by other dams downstream from Three Gorges, making the world's largest powerhouse superfluous (China Online 2003). Chinese dissidents have suggested that massive corruption has led to shoddy construction, with increased likelihood of dam failure (China Online 2003; Probe International 2003). Resistance from local citizens to resettlement also continues to cause international human rights groups to criticize the government and the project (International Rivers Network 2003b).

It is worth noting that the purposes of the Three Gorges Dam as the government expresses them—as well as much of the criticism leveled at it—map almost precisely onto the dam projects from the mid-twentieth century U.S. With the Three Gorges Dam, China is ramping up its own growth machine: producing more electricity than it currently has use for, expanding transportation routes that may be useless in less than a generation's time, and expanding irrigation and drinking water facilities for a growing population, all for the unimaginable cost of $24-$75 Billion and at the risk of a massive tragedy if the dam were indeed to fail.

These dams show that in quite varied circumstances—socialist and neo-liberal, democratic and authoritarian, industrial capitalist and post-colonial nations—the treadmill of production is alive and well. Nations are using the model of the U.S. West, including the technology and occasionally the same construction companies, to mount a treadmill of production to spur development: energy production, agricultural production, drinking water for booming cities, and flood control to protect an expanding population; in short, for overall economic growth. The fact that the U.S. West is facing serious problems with continued expansion has not stopped new development elsewhere. Perhaps it is the ideological denial of any problems by the U.S. planners and politicians that encourages developing nations.

These dams, however, do not necessarily represent on-ramps to a treadmill of production for individual nations. Development in the context of globalization, with hegemonic U.S. power, has continually shown that nations of the global South will continue to remain dependent in many ways, regardless of their own internal attempts at establishing a treadmill of production on the U.S. model (Gould and Schnaiberg 2000). Instead, then, it is likely that such attempts by developing nations will contribute to a *global* treadmill of production, with the global South continuing to supply ecological and labor resources to produce for the consumption of the U.S. and the West.

But the dams under construction have seen strong opposition, both locally and internationally, and this also shows the lessons learned from the earlier environmentalist movements that gained some success in the mid-twentieth century U.S. Mobilized by local histories of struggle as well as international support, encouraged by celebrity activists and wide attention thanks to the quick spread of information via the internet, many local struggles have articulated a vision of social justice, cultural and historical preservation, and indigenous and subaltern class rights, all linked to varied forms of environmentalism. Though the typical methods of intimidation, buy-offs, and the exercise of sheer political power have often been successful in neutralizing this opposition, the mere fact of continual struggle in so many locations offers hope of slowing the global treadmill.

UNINTENDED NEGATIVE CONSEQUENCES

The process of state building in the American West thus shows a diversity of experiences and results. Scott (1998) has argued that many processes of state building lead to failure when they exhibit a certain set of conditions. In particular, a strong authoritarian state with an ideology of high modernism and a weak civil society leads to a "failure to improve the human condition." Though he does not identify the American West as such an example, the implication of his work is that elements of high modernism must have been implicit at the very least in the Progressive-Era and New Deal U.S. The historical work of this book has shown how a vibrant civil society has both called upon the state to build an infrastructure in the West, and at other times stopped the state from doing so. This variation shows a deep complexity in history; the ideology of high modernism, as it is exhibited in these cases without an anemic civil society, did not lead to utter failure, as Scott's examples did.

This work has also shown the ways that the state-building process has produced environmental change, both through imposing a physical infrastructure as well as enforcing a nature-society relationship—one characterized by domination. The ways that the Progressive-Era and New Deal state imposed itself on the landscape have had myriad consequences, both beneficial and destructive. Many of the goals of state building have been achieved, including the construction of a dynamic society and growing economy. The collection of drinking and irrigation water, the production of hydropower, and the prevention of flooding have all provided a foundation for the building of an American empire in the West—both James Rorty in the 1930s and Hunter Thompson in the 1970s have called it the fulfillment of the American Dream. This fulfillment, however, is not without its unintended—and negative—consequences.

The most obvious of the destructive consequences from the many dams in the American West are the impacts on river ecosystems and fish and wildlife. When a dam is placed on a river, normal flows are stopped and many miles of upstream riverbed are transformed into lake. The dams, however, also produced unforeseen and unintended consequences. The deep reservoirs of the West hold very cold water, which when released through the hydropower plant, yield an altered temperature and ecosystem downstream as well. The impact on ecosystems is well documented—plant and wildlife populations and geologic formations are all affected (American Rivers, Friends of the Earth and Trout Unlimited 1999; Glen Canyon Action Network 2001; Worster 1985). The plight of salmon in the Pacific Northwest is well known (one sub-species, the Snake Basin Coho was declared extinct in 1985, wholly due to dams), and recently the cases of the pike minnow, humpback chub, razorback sucker, snail darter, suckerfish, and other such endangered species have become important symbols for environmentalists. These cases and many others stem from the damming of the rivers of the West. But beyond the effects on animals and the health of the river ecosystems, the unintended consequences for society are great as well.

As the dams of the West silt up and become less effective, society will have to address problems of water scarcity and a reduced hydropower production at exactly the times when governmental action is needed the most. The population of the West continues to grow, and the infrastructure put in place in the middle of the twentieth century will be able to provide less and less of society's needs. Interestingly, as this crisis looms on the horizon in 2002, the federal government is actually reducing its capacity to address these and other problems. The Bureau of Reclamation has initiated a program of "title transfers," which marks a new phase of reclamation in the West. Title transfers, as discussed in chapter three, absolve the federal government of responsibility for water projects such as dams, canals, and reservoirs, and grants ownership to local and state groups. This reduced state capacity means that when crises hit, the national state will have less ability to respond meaningfully to water scarcity or other problems. Privatization of water systems worldwide has often ended in failure (Finnegan 2002), and the contemporary U.S. is not likely to be an exception.

The Bureau of Reclamation has officially recognized that it cannot build any new high dams in the West. As one report states, "the demand for water continues to increase as the supply decreases" (United States Bureau of Reclamation 1996). Furthermore, "the justification for . . . building new facilities has, by and large, gone away" (Hess 1996). The sources of constraint upon dam building are many, not the least of which are the natural limits. All but a few major rivers in the American West are dammed (most of them multiple times). In the United States there are some 75,000 dams; in California alone there are some 1,200 major water projects. California stands as an example of how the U.S. has become a dammed nation.

The discourses of imperial high modernism, though sometimes mobilizing arguments based on history, effectively erase much of the contestation and opposition to the state's plans. The discourse of improved nature, though sometimes glorifying particular elements of history (reclaiming marginal land), similarly selects some aspects of a narrative and forgets about others. This apparent historical amnesia on the part of the federal government shows an inability to learn from past mistakes. The historical examinations of this book have shown how the elision of facts, a lack of opposition, and the rampant developmentalist ideology of high modernism led to the current unsustainable state of affairs. Though the dams of the American West are by no means failures on the level of Scott's cases, the potential for failure remains. As we search for solutions to the inevitable water crisis in the West, we must remember that no one spoke for the Columbia River salmon; that Glen Canyon was the 'place no one knew'; that left to its own devices, nature may act back in ways that undermine human attempts to capitalize on natural 'resources.'

A CIVILIZED WILDERNESS?

State building, as this work has shown, is largely a process of civilizing nature—both human nature and non-human nature. In this process, exemplified by the damming of the rivers of the American West, nature and society were constructed in a double fashion. Rivers were seen simultaneously as wild, raging torrents and "menaces" to society; they were also seen as resources that were "wasted" by running to the sea untapped. This double construction as passive agent (a wasted stream) and active force (menace to society) shows that the nature-society relationship is conditioned more by social contradictions than it is by any objective reality inherent in the rivers of the American West. The rivers are given the meanings that they carry by humans eager to further their own social agenda. Governments want to harness resources to spur growth and productivity. Local elites want to control one of the most important foundations of society in an arid region. Environmentalists want to preserve a scenic landscape for recreation and posterity. Each social group discursively constructs nature to take the form that supports their social agenda. For the government, the streams are being wasted, and instead should be harnessed to support industry and farming to build an empire. For local elites, they are similarly wasted, and should be used wisely to benefit the local people and economy. For the environmentalists, the rivers are a sacred symbol of wildness and freedom: to dam them is symbolically to subdue the landscape of the West, robbing its vitality.

Nature, including what environmentalists have called wilderness, has been dominated in the process of state building, and one set of outcomes is the creation of recreation areas. In these areas, each group has attempted to define nature discursively to benefit itself or its cause. The Sierra Club calls Glen Canyon Dam a desecration of sublime Nature while the Bureau of Reclamation views the dam as an improvement in access to the sublime. Nature has been subjected to a civilizing process, one in which much of the wildness has been removed.

The trouble with wilderness, though, is that it is constructed in such a way that human entry into it desecrates it by definition (Cronon 1996: 83). If wilderness is that area not inhabited by humans, then even our direct appreciation of it is an intrusion. The concept of nature as a garden (Pollan 1991), or an area with and without human inhabitants as containing similar wildness, dissolves the binary of nature versus society. Indeed, the weight of research in environmental history has shown that the places humans consider most wild were in fact strongly marked by human passage.

Wallace Stegner noted this in many of his works, including an essay in the magazine *Holiday* in 1966 entitled "Glen Canyon submersus" where he commented on the loss of Glen Canyon:

> And yet, vast and beautiful as it is, open now to anyone with a boat or money to rent one . . . democratically accessible and with its most secret beauties captured on color transparencies at infallible exposures, it strikes me, even in my exhilaration, with the consciousness of loss. In gaining the lovely and the usable, we have given up the incomparable.

Stegner recognized the double-sided construction of nature, both in his work defending Dinosaur National Monument and in mourning the loss of Glen Canyon. He was sensitive to the human history involved with areas denoted as wilderness, the "marks of human passage" as he called them. As this effort has shown, dams symbolize the marking of human passage on a monumental level—a civilizing of the wilderness that both transforms a "menace to a resource," "descerates" a "miracle," and also provides "democratic access" to places of natural beauty.

What this shows us is that the discourses of high modernism and environmentalism are real—they are not just texts to be read, interpreted, and deconstructed. Discourses and ideas matter as they are connected to material power; The ways that people and institutions construct reality through discourse have had real effects on the landscape of the West. Indeed, they have had a real impact on people's lives, their livelihood, and even their bodies. The high modernist discourses that justified the building of the dams helped to produce the metropoles of Los Angeles and Las Vegas. They helped California grow to prominence in the United States' economics, politics, as well as culture.

Without water and hydropower, California's society would not have the shape that it now occupies. The United States would have a very different west coast, and without a large western population, the national political landscape would have a very different topography. Water politics in the West thus holds a central place in U.S. history for many reasons. The imperial high modernist discourses legitimating state building helped to bring this about, and the environmentalist discourse helped to put the brakes on this blind development.

This shift in the valuation of nature points to a larger transformation of the relationship between nature and society, a transformation that is in process, unevenly completed, and perhaps will never be as hegemonic as high modernist was. But this new view of the nature-society relationship—based on preservation and inherent natural value—has destabilized the hegemonic high modernist ideology. In offering an effectual foundation for oppositional discourse, the environmentalist ideology has stopped the damming of the rivers of the West, and forced society to re-examine its relationship to growth, natural resources, and state building.

But the beginning of a shift in the relationship between society and nature is just that: a beginning. The oppositional discourse of preserving aesthetic nature and valuing nature for itself is a powerful one. It has brought the environmental movement in the U.S. to the bargaining table of politics, and important victories have been won. Yet the Bureau of Reclamation and other political agencies have reacted to this discourse in immediate and substantial ways. The cooptation of this environmentalist discourse may lead to the undoing of the environmentalist movement in the future. If the "agents of blind progress," as the Sierra Club once called the Bureau of Reclamation, successfully coopt the rhetoric of aesthetic value in nature, environmentalists could find their own discourse opposing them. The case presented in chapter six was one of the first examples of the use of a new discourse of improved nature. The beauty and accessibility of Lake Powell were asserted as justification for the dam at Glen Canyon. This is potentially grounds for the resurgence of an ideology of growth and domination.

The discourse of improved nature presumes that the increases in democratic access to nature trump any loss of pristine wilderness. This shift in discourse posits an improvement on the previously existing natural conditions due to an extraction of resources and an improvement in availability. This shift has the potential to undo the effects of the oppositional discourse that constructs an inherent value of nature in a wilderness state. The appeal to democratic access draws on a long history of cherishing liberty and equality that, as de Tocqueville (1840/1990) argued, can oppose one another and produce a leveling effect. If Americans, in their search for democracy and equality in access to nature are willing to degrade natural areas, then we are in danger of legislating to the lowest common denominator in appreciation and value of nature.

Nearly 70 years ago Aldo Leopold commented on "The Virgin Southwest," quoted in the epigram to this chapter. One of his ideas for a solution to controlling the engines of progress was to apply science and ethics to land-use, a solution that his farm in Wisconsin still exemplifies. The other solution was to "cultivate a love of country a little less spangled with stars, and a little more imbued with . . . respect for mother-earth." Indeed, the star-spangled love of country (such as the imperialistic high modernist ideology) needs to be tempered with a respect for "mother earth" and the aesthetic beauty and value therein (the environmentalist discourse). It is tempting to search for a middle ground between these two contradictory discourses where state action might be more sustainable, environmentally minded, and less destructive. But the momentum that our destructive practices have achieved combined with the cooptation of opposition means that such a middle ground will be extremely difficult to find and to maintain. It is, nonetheless, our only hope.

Maps

BOULDER (HOOVER) DAM AND THE LOWER COLORADO RIVER

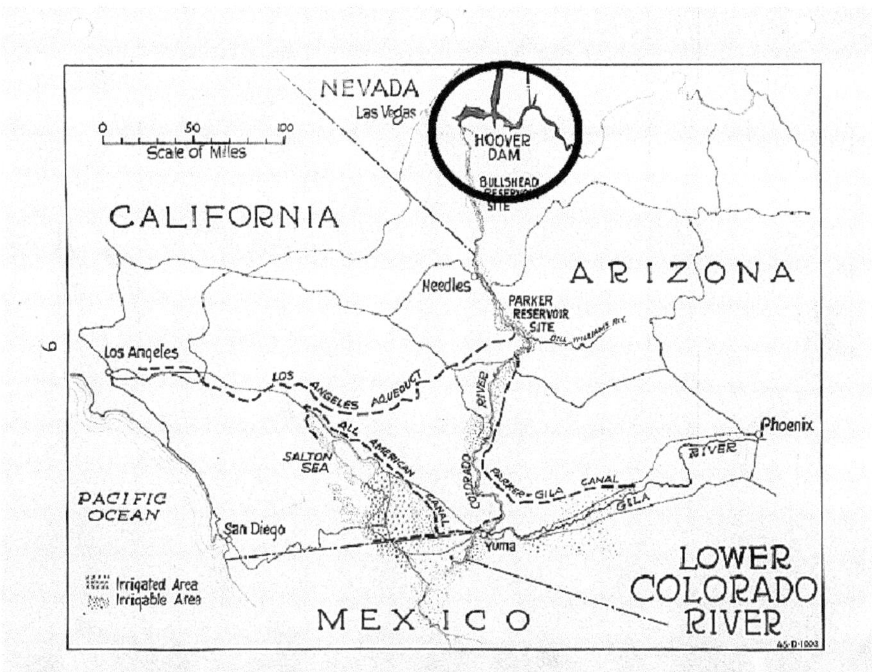

Map courtesy of the Bureau of Reclamation

GRAND COULEE AND THE COLUMBIA BASIN

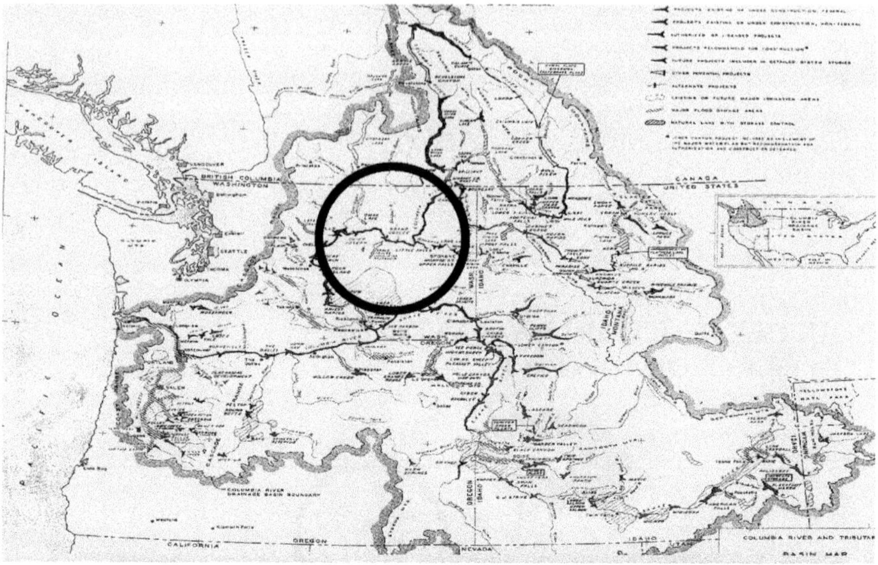

Map courtesy of the Bureau of Reclamation

GLEN CANYON AND THE UPPER COLORADO RIVER

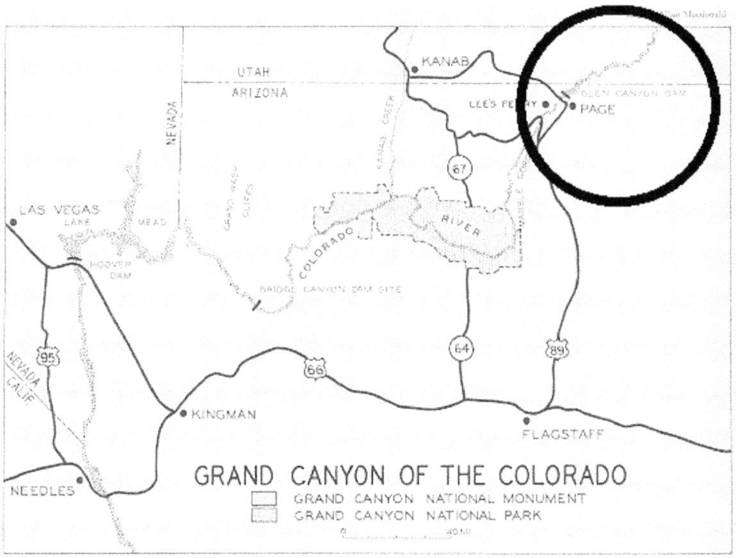

Map courtesy of the Bureau of Reclamation

Notes

NOTES TO CHAPTER ONE

1. Throughout this work I will use the term natural to indicate non-human processes, entities, or characteristics. I counterpose this to the term social, which involves human endeavor. As I will discuss, however, these two realms are thoroughly imbricated in one another through historical actions as well as through recollection and reconstruction. Thus, natural formations such as climate, geography, or soil composition should be differentiated from obvious social products such as the built environment of cities, but also from non-obvious but non-trivial human manipulations such as plant breeding, species transportation, and etc. For example what is normally construed as a "natural disaster" such as a flood destroying a town is more appropriately termed a social disaster, for it was a social decision to place a community in a flood plain. The natural and the social exist in a dialectical embrace of mutual constitution and transformation.
2. James Scott's work, *Seeing Like a State,* argues that high modernism is a world view in which the "strong version of the beliefs in scientific and technical progress that were associated with industrialization in Western Europe and in North America from roughly 1830 until World War I" was transcendent. Scott defines high modernism as a "supreme self-confidence about continued linear progress, the development of scientific and technical knowledge, the expansion of production, the rational design of social order, the growing satisfaction of human needs, and, not least, an increasing control over nature (including human nature) commensurate with scientific understanding of natural laws" (1998: 89). This theoretical perspective will be more fully explored in chapter two.
3. A similar set of arguments is made by John Wesley Powell in his famous identification of the beginning of the West at the 100th meridian (Powell 1878).
4. Boosters used explicit metaphors describing the area around a given project as a "Promised Land," an "Eden," or a "paradise." Though it may seem paradoxical to the modern reader that any such Land of Canaan would need to be completed by the state, such was their argument (see chapter five). In fact, in some cases, boosters had to scale back their grandiose rhetoric about the Edenic qualities of the West because Eastern investors became unwilling to back irrigation projects, thinking that improvements were superfluous (Farr 1918).

NOTES TO CHAPTER TWO

1. This term is Richard White's, *The Organic Machine* (1995).
2. The analytical separation of humans from nature, or society from environment is a major concern of this work. Much of this chapter examines this socially constructed binary, concluding that it makes little sense to posit a hard-and-fast boundary between society and nature. Instead, we might take a historical perspective on the dialectical interactions between society and nature, understanding that the two concepts are so thoroughly imbricated in one another that to speak of them as separate, even for analytical purposes, can be misleading.
3. Examples of such crises abound. A historical example is the disappearance of the Anasazi civilization in what is now the southwestern United States (Bullock 1991; Turner and Turner 1990). Present examples are found in Bill McKibben's *The End of Nature* (1990), or Mike Davis' critique of the development of the Los Angeles Basin in *The Ecology of Fear* (1998).
4. Schnaiberg pays attention to this era when he examines the ways in which the state has redistributed revenue to labor: "So the state is also involved with giving. First the state employs workers to provide services to capital and labor. In one sense this is a keeping of revenues. But the wages thereby generated flow out of the state and into markets. Second, the state contracts with private industry to build capital projects, such as dams and highways, again releasing funds from the state into markets. Moreover, such physical capital enables private sector investments to be made, since profits are possible when large-scale infrastructure has been provided. Third, the state redistributes its revenues as transfer payments, generally to various categories of labor" (Schnaiberg 1980: 242).
5. When discussing the state, it is all too easy to fall into obscurantist language. A state is, after all, made up of both individual actors and individual agencies. While I do not want to intervene in the debate over methodological individualism, I do wish to acknowledge that both structure and agency influence the complex ways that the state makes a decision. Furthermore, the use of the definite article *the* when describing the state also tends to conceal the profound fragmentation, conflict, and contradictions within different branches, bureaus, or administrations of a single state.
6. I use a definition of the state from Weber: "The primary formal characteristics of the modern state are as follows: it possesses an administrative and legal order subject to change by legislation, to which the organized activities of the administrative staff, which are also controlled by regulations, are oriented. This system of order claims binding authority, not only over the members of the state, the citizens, . . . but also to a very large extent over all action taking place in the area of its jurisdiction. It is a compulsory organization with a territorial basis. Furthermore, today, the use of force is regarded as legitimate only so far as it is either permitted by the state or prescribed by it. . . . The claim of the modern state to monopolize the use of force is essential to it as a character of compulsory jurisdiction and of continuous operation" (Weber 1922/1978: 56).

7. It must be made clear that Scott does not argue that the twentieth-century American West is an example of high modernism. Scott's examples include several wide-ranging cases: the building of a new capital for Brazil at Brasilia, compulsory villagization in Tanzania, and forced collectivization in the Soviet Union. A chapter written about the Tennessee Valley Authority was left out of the final version of the book due to space considerations. The assumption in this book is that a geographically and historically punctuated ideology makes little sense, especially since Progressive-Era North America is used as a benchmark in Scott's definition (Scott 1998: 89). In between the places and times of Scott's cases, the ideology of high modernism must have been immanent at the very least. The case of the American West presents a counter example to a monolithic high modernism.
8. Here I am clearly relying on a normative base for the definition of *sustainable* and *intrusive*. While we may argue (often *ad nauseum*) over whether a certain set of practices are more or less sustainable (see page 14), there is a point where we can assume, Julian Simon (1981) notwithstanding, that resources will be exhausted.
9. These authors represent only a few of the many that could be cited on this subject. To continue this partial list, see Crosby (1986), Marsh (1970/1864), Merchant (1989), Pyne (1990), and Webb (1931/1959).
10. There is, of course, a large counter-current to this ideology, as exemplified by John Muir's Sierra Club and the fight against Hetch Hetchy Dam (see chapter three).
11. This was not the first time that nature was seen as a place of escape to relieve the tensions of the city. Through the long history of western ideas of nature this idea was first expressed in the mid-nineteenth century by authors such as Thoreau. The metaphor of escape was occasionally resurrected at different times through the nineteenth and twentieth centuries by figures such as Frederick Law Olmstead and John Muir (see Cronon 1996, Nash 1967/1982, and Worster 1994).
12. The rise of the middle class cannot be seen as the only cause of mid-twentieth century environmentalism, but it is likely the most important. Though the rise of environmentalism in the mid- to late twentieth century is correlated with the rise of the middle class, it is also correlated with the rise of neo-conservatism and anti-environmental political movements that support reduced governmental regulation, thus giving a wrinkle of complexity to an otherwise tempting causal argument. Some scholars have also used environmentalism as an example of new social movements (Melucci 1996), an example of post-materialist politics (Inglehart 1977), and as an entirely new paradigm (Catton and Dunlap 1978; Martinez-Alier 1995).

NOTES TO CHAPTER FOUR

1. Black Canyon is the name of the site where Boulder Dam was built. The first name for the dam is after the original site, Boulder Canyon, slightly upstream from Black Canyon. In his 1922 report to Congress, Arthur Powell Davis, Commissioner of Reclamation, suggested a dam "at or near the vicinity of

Boulder Canyon" (Bureau of Reclamation 1922: 21). Thus in 1928 Congress passed the Boulder Canyon Act, and the project was officially named after Boulder Canyon even though Black Canyon was ultimately chosen as a more appropriate site for the high dam.

2. Due to the changing political-economic structure of society, many of these dams are now being removed. The state of Wisconsin is leading the way in this process. By 2003, various state and non-governmental groups have allied to remove 73 of the approximately 3,700 dams in the state. Approximately 485 dams have been removed of 75,000 total dams that have been built nationwide (American Rivers, Friends of the Earth, and Trout Unlimited 1999: viii; American Rivers 2003; Wisconsin Stewardship Network 2003).
3. The confusion over the dam's name arises largely from a partisan political squabble. Because Herbert Hoover played such a large role in brokering the Colorado River Compact, President Hoover's Secretary of Interior Ray Lyman Wilbur, at the groundbreaking for the railroad supply line to the dam site on September 17, 1930, made the announcement that work had just begun on "Hoover Dam" (Stevens 1988: 33). Most boosters, however, continued to call the project after the legislation: the Boulder Canyon Act. As fate would have it, the dam was finished in 1936 during Franklin D. Roosevelt's first term of office, and so Roosevelt officially christened the dam as Boulder. The present name, Hoover Dam, was given by a 1947 act of Congress.
4. President Roosevelt held a dedication ceremony in September of 1935 officially naming and opening the dam, though final construction was not completed until 1936.
5. This cost total is strictly monetary—including the social costs in the price of the dam would have to account for the deaths of at least 110 workers.
6. This archaic word bears some explanation: *Omni:* all, *-parous:* to bring forth, to bear. Omniparous, in this context, means that which brings forth all variety of crops.
7. This is, of course, only one interpretation of the role of nature in a Christian theology. Bill McKibben's reading of the Book of Job concludes that God's mystery is not to be understood by man, including mysteries such as why the rain falls in unpopulated areas (McKibben 1990: 70). This interpretation highlights a radical anti-anthropocentrism that is atypical of many forms of Christian faith.
8. Manuscript held in the Phil D. Swing papers, UCLA Special Collections Swing 1890–1963.
9. Manuscript held in the Harold L. Ickes Papers, box 118, Library of Congress.
10. In the interpretation offered here, Swing's comments do not represent a pacifist, anti-interventionist, or anti-war position; rather a streamlining or governmental efficiency argument that the war effort is better served by the production of power.
11. The use of such language supports the Elizabeth Sanders' (1999) arguments about the continuity of effort and political alliance between Populist Era farmers and Progressive Era reforms.
12. The Teapot Dome scandal involved the private appropriation of oil from public lands. President Taft had created naval oil reserves in California and

President Wilson enlarged this by adding the Teapot Dome reserves in Wyoming. The scandal came about in 1921 when President Harding transferred the administration of the naval oil reserves from the Secretary of the Navy to the Department of Interior, where Secretary Albert B. Fall, "whose past political and business experience had already stamped him as one of the deadliest enemies of the conservation movement" gave permission to oil companies to mine the reserves for a bribe of $400,000 (Robbins 1976: 399).

13. This galley version of the pamphlet, held in the Phil Swing Papers at UCLA shows Swing's comments. It does not have a publication date, but can be estimated as 1923.
14. Letters of affiliation held in the Phil Swing Papers, UCLA.
15. Letter held in folder 10 of the Charles Squires Papers, University of Nevada, Reno.
16. Manuscript held in the Harold L. Ickes Papers, box 118, Library of Congress.
17. Press release copy with Swing's notations held in the Phil Swing Papers, UCLA.
18. Manuscript held in the Harold L. Ickes Papers, box 413, Library of Congress.
19. This is one of the few instances of a booster calling the dam by this name.
20. This pamphlet, held in the Swing Papers at UCLA carries no author or publishing date. Swing's writing on the outside gives the date and attributes authorship to Harry Chandler, who did sign the introduction. Chandler will, for simplicity, be identified as the sole author of what was likely a collaborative work.
21. According to handwritten notes on newspaper clippings and pamphlets in Swing's scrapbooks, held in the Swing Papers at UCLA, Swing cautiously identified the distributor of the pamphlets as Chandler.
22. During the campaign, Chandler's *L.A. Times* was not above outright libel. The *L.A. Times* supported Ed Sample, and according to the San Diego *Union* and the Fullerton *News* 25 August 1924), Chandler and his associates distributed a flyer announcing that the Ku Klux Klan endorsed Swing. The *News* article quoted Swing: "'It was expected that the *Times* would make some eleventh hour move against me in a desperate effort to divert the minds of the voters for the district from a tremendous issue in my campaign. The issue to which I refer is the issue of Boulder Dam and the All-American Canal. If I am defeated, Harry Chandler of the *Times* and his associates will make fifty or sixty million dollars. They plan my defeat so that the All-American Canal may be eliminated and the American waters in the Colorado River conserved by American dollars will be allowed to flow on Mexican land owned by Harry Chandler. Shall this priceless heritage of water be allowed to slip away from America?"
23. Manuscript held in the McCluskey papers, University of Arizona.

NOTES TO CHAPTER FIVE

1. An acre-foot is the measurement of water used in agricultural and reclamation publications. It is the amount of water required to cover one acre of ground one foot deep (approximately 326,000 gallons, or 1.3 million liters).

2. Pitzer (1994) makes the argument that Grand Coulee power was not exactly *crucial,* but instead was simply *helpful.* He rightly asserts that the aluminum industry may have been attracted to the Pacific Northwest by the lure of cheap power, but that the wartime power needs could have been met had Grand Coulee not come on line. The Grand Coulee hydropower was helpful in that it did not force other consumers to curtail their use, but it was not crucial, as sufficient power could have been appropriated from other sources.
3. This debate connects with the national discussions of public power production and distribution, especially as public power was instituted at the Tennessee Valley Authority (Ellison 1999).
4. James Scott argues that discourses of high modernism have a "temporal emphasis . . . almost exclusively on the future" and tend to view "the past [a]s an impediment" (1998: 95). While the idea of linear progress is certainly a main tenet of high modernism, which is not contradicted by the Grand Coulee case, boosters also tended to use any available persuasive arguments, including those that relied on recent history for legitimacy.
5. The pamphlet has no official publication date but it quotes statements from 1928 by both President Coolidge and Secretary of Commerce Herbert Hoover, thus establishing it as being published before Hoover took office in early 1929.
6. The book has no official publication date but due to the authorial style and the description of the dam site, University of Washington Librarians date the pamphlet to the mid-1930's.
7. Letter held in the National Archive and Records Administration, Washington, D.C., Record Group 48, Central Classified File 1907–1936, Entry 8–3 Colorado River and Columbia Basin, Box 1592.
8. Letters to and from Powell, Mead, and Secretary of the Interior Hubert Work from 1923 through 1928 show the leaders to be concerned about the efficient use of government funds. The letters are held in the National Archives and Records Administration, Record Group 48, Entry 765, Box 3, Bureau of reclamation Files. Powell, and later Mead, came to believe that a large, successful project (specifically one at Boulder Canyon) could save the Bureau by granting it a legitimacy that had been lost in the early 1920s (see chapter four).
9. Press releases are held in the National Archives and Record Administration, Record Group 48, Entry 7, Projects Correspondence, Box 477.
10. As quoted in Nixon 1972: 40.
11. As quoted in Nixon 1972: 42.
12. As quoted in Nixon 1972: 40.
13. Manuscript held in the Harold L. Ickes Papers, box 414, Library of Congress.
14. Guthrie was paid at the rate of an annual salary of $3,500 for his month of labor, which meant a total of $266.66, or about $10 per song. In 1945 when the BPA sent him the original scores that they had retained, they requested that at each performance of one of the songs the BPA be acknowledged, which Guthrie dutifully did. Interestingly, Guthrie had to sign an oath of office swearing to uphold and defend the U.S. Constitution as a condition of his employment (Murlin 1991: 11, 88).

15. Guthrie's widespread popularity was largely posthumous. By the mid-1960s the Guthrie estate was earning $50,000 per year in royalties, in part from the use of his songs by other rising stars of the folk tradition such as Bob Dylan, Tom Paxton, and Joan Baez (Klein 1980: 433).

NOTES TO CHAPTER SIX

1. These dams were proposed as two structures with a single purpose. I follow the shorthand that is common in the general historical literature, and call this unit "Echo Park," and thus the debate that centered on the submersion of parts of Dinosaur National Monument is usually dubbed the "Echo Park controversy" (see, for example, Fox 1981; Gottlieb 1988; Nash 1967/1982).
2. Letters and pamphlets held in the Sierra Club Collection, Bancroft Library, University of California, Berkeley.
3. The idea of the river as an "economic resource" as opposed to a "natural" and "aesthetic" resource is a false dichotomy. Activists constructed this binary so as to fight the economic logic of using the river for society's ends. But what they did not recognize (or chose not to make explicit) was that the aesthetic use of the river is just as anthropocentric as an economic use, though it may be more sustainable. Activists did use some rhetoric about the qualities of the river in and of itself. This was largely understood to be a benefit for humanity in terms of recreation, spiritual regeneration, or simply aesthetic pleasure. As Cronon has shown, wilderness is a human creation, a mirror that reflects "our own unexamined longings and desires" (1996: 70). These ideas will be revisited in the concluding chapter, with an example of a discourse of "improved nature." This discourse explicitly masks the past economic use of nature by employing the tropes of the aesthetic re-creation benefits of nature.
4. All letters quoted in this section are held by the National Archives and Record Administration, Record Group 48, Central Classified Files, entry number 4–4; boxes 360–64.
5. The text of this response letter is held by the National Archives and Record Administration, Record Group 48, Central Classified Files, entry number 4–4; box 364.
6. This memorandum is held by the National Archives and Record Administration, Record Group 48, Central Classified Files, entry number 4–4; box 363.
7. Memorandum held at the National Archive and Records Administration, Record Group 48, Central Classified Files, entry number 4–4; box 363.
8. Diaries held in the David Brower archives, Bancroft Library, University of California, Berkeley.

NOTES TO CHAPTER SEVEN

1. Leavitt's statement is characteristic of imperial high modernism, but this particular quotation lacks a clear statement on the means of achieving the "orderly development of population and resources," which would include the rational application of science and technology.

Bibliography

Abbey, Edward. 1968. *Desert Solitaire: A Season in the Wilderness.* New York: Ballantine Books.

Abbey, Edward. 1975. *The Monkey Wrench Gang.* Philadelphia: Lippincott.

Abbey, Edward. 1990. *Hayduke Lives!* New York: Little, Brown and Co.

Abbott, Carl. 1994. "The Federal Presence." Pp. 468–499 in *Oxford History of the American West,* edited by C. A. Milner, C. O'Connor, and M. A. Sandweiss. Oxford: Oxford University Press.

Alexander, J. A. 1928. *The Life of George Chaffey, a Story of Irrigation Beginnings in California and Australia.* London: MacMillan.

Alexander, Thomas. 1960. "John Wesley Powell, the Irrigation Survey, and the Inauguration of the Second Phase of Irrigation Development in Utah." *Utah Historical Quarterly* 37:190–206.

American Rivers. 2003. http://www.amrivers.org/damremoval/default.htm. Accessed 2003.

American Rivers, Friends of the Earth, and Trout Unlimited. 1999. "Dam Removal Success Stories: Restoring Rivers Through Selective Removal of Dams that Don't Make Sense." American Rivers, Friends of the Earth, Trout Unlimited.

American Rivers, Friends of the Earth, and Trout Unlimited. 1999. http://www.tu.org/scripts/trout/newsletters/ "Dam Removal Success Stories: Restoring Rivers Through Selective Removal of Dams that Don't Make Sense."

Anderson, Benedict. 1991. *Imagined Communities.* London: Verso Press.

Anderson, Perry. 1974. *Lineages of the Absolutist State.* London: Verso.

Baxandall, Rosalynn and Linda Gordon. 1995. *America's Working Women: A Documentary History 1600 to the Present.* New York: Norton.

Berger, Peter and Thomas Luckman. 1967. *The Social Construction of Reality.* Garden City: Anchor Books.

Bohn, Frank. 1927a. *The Boulder Canyon Dam: The Essence of the Swing-Johnson Bill.* New York: Joint Committee of National Utility Associations.

Bohn, Frank. 1927b. *Boulder Dam: From the Origin of the Idea to the Swing-Johnson Bill.* New York: Joint Committee of National Utility Associations.

Boulder Dam Association. 1923. "Boulder Dam." San Diego: Boulder Dam Association.

Boulder Dam Association. 1926. "Boulder Dam Project Should Not Be Delayed By Arizona." Los Angeles: Boulder Dam Association.

Boulder Dam Association. 1927a. "Bulletin No. 101." Los Angeles: Boulder Dam Association.

Boulder Dam Association. 1927b. "Bulletin No. 102." Los Angeles: Boulder Dam Association.

Boulder Dam Association. 1927c. "Bulletin No. 103." Los Angeles: Boulder Dam Association.

Boulder Dam Association. 1927d. *The Federal Government's Colorado River Project.* Los Angeles: Boulder Dam Association.

Braverman, Harry. 1974. *Labor and Monopoly Capitol: The Degradation of Work in the Twentieth Century.* New York: Monthly Review Press.

Brown, Wendy. 1988. *Manhood and Politics: a Feminist Reading in Political Theory.* Totowa, New Jersey: Rowman and Littlefield.

Brulle, Robert J. 2000. *Agency, Democracy, and Nature: The U.S. Environmental Movement from a Critical Theory Perspective.* Cambridge: MIT Press.

Buenker, John D. 1973. *Urban Liberalism and Progressive Reform.* New York: Scribner's.

Bullock, Peter Y. 1991. "A Reappraisal of Anasazi Cannibalism." *Kiva* 57:5–16.

Burawoy, Michael. 1982. *Manufacturing Consent: Changes in the Labor Process under Monopoly Capitalism.* Chicago: University of Chicago Press.

Burawoy, Michael and Theda Skocpol. 1982. *Marxist Inquiries: Studies of Labor, Class, and States.* Chicago: University of Chicago Press.

Bureau of Reclamation. 1922. *Development of the Imperial Valley (The Fall-Davis Report).* Washington, D.C.: Government Printing Office.

Bureau of Reclamation. 1932. "Illustrated Lecture on Boulder Canyon Project." edited by Office of the Chief Engineer. Boulder City: Department of the Interior Bureau of Reclamation.

Bureau of Reclamation. 1933. "Boulder Dam." Boulder City: United States Department of Interior Bureau of Reclamation.

Bureau of Reclamation. 1936. "Boulder Dam: The World's Most Impressive Engineering Spectacle in the Center of an Area of Unsurpassed Scenic and Historical Interest." Boulder City: United States Department of the Interior Bureau of Reclamation.

Bureau of Reclamation. 1938. *Grand Coulee Dam: The Columbia Basin Reclamation Project.* Washington, D.C.: United States Department of the Interior.

Bureau of Reclamation. 1946. *The Colorado River: A Comprehensive Report on the Development of the Water Resources of the Colorado River Basin for Irrigation, Power Production, and Other Beneficial Uses in Arizona, California, Colorado, Nevada, New Mexico, Utah, and Wyoming.* Washington, D.C.: United States Department of the Interior.

Bureau of Reclamation. 1961. *Reclamation Project Data.* Washington, D.C.: Government Printing Office.

Bureau of Reclamation. 1964. *The Story of the Columbia Basin Project.* Washington, D. C.: Government Printing Office.

Bureau of Reclamation. 1965. *Lake Powell: Jewel of the Colorado.* Washington, D.C.: Government Printing Office.

Bureau of Reclamation All American Canal Board. 1919. *All-American Canal Report (Elwood Mead, W. W. Schlecht, G. E. Grunsky).* San Francisco: Bureau of Reclamation.

Burrows, Malcolm. 1996. *The Neurobiology of an Insect Brain.* Oxford: Oxford University Press.

California Legislative Analyst's Office. 2000. *California's Economy and Budget in Perspective*. Sacremento, CaLAD.

Case, Robert Ormond. 1935. "The Eighth World Wonder." *Saturday Evening Post,* July 13, 1935.

Catton, William R. JR and Riley E. Dunlap. 1978 "Environmental Sociology: A New Paradigm." *American Sociologist*. 13(1): 41–49.

China Online. 2003. "Three Gorges Dam Project." http://www.chinaonline.com/refer/ministry_profiles/threegorgesdam.asp. Accessed October 2003: China Online: The Information Network for China.

Chomsky, Noam. 1957. *Syntactic Structures*. The Hague: Mouton.

Clements, Kendrick. 1979. "Politics and the Park: San Francisco's Fight for Hetch Hetchy, 1908–1913." *Pacific Historical Review* 48:185–215.

Columbia Basin Commission. 193? *How to Use the Grand Coulee Dam to Advertise the Resources and Recreational Assets of the State of Washington*. Olympia, WA: State of Washington.

Columbia Basin Commission. 1928. "The Columbia Basin Irrigation Project: National economics Justify Columbia Basin Project." Pp. 16. Olympia: Columbia Basin Commission.

Confederated Tribes of the Colville Reservation. 1994. *Grand Coulee Settlement*. Nespelem, WA: Confederated Tribes of the Colville Reservation.

Coronil, Fernando. 2001. "Smelling Like a Market." *American Historical Review* 106.

Cronon, William. 1983. *Changes in the Land: Indians, Colonists, and the Ecology of New England*. New York: Hill and Wang.

Cronon, William. 1991. *Nature's Metropolis: Chicago and the Great West*. New York: Norton.

Cronon, William. 1994. "Landscapes of Abundance and Scarcity." Pp. 602–637 in *Oxford History of the American West,* edited by C. A. Milner, C. A. O'Connor, and S. Martha A. Oxford: Oxford University Press.

Cronon, William. 1996. "The Trouble with Wilderness; or, Getting Back to the Wrong Nature." in *Uncommon Ground: Rethinking the Human Place in Nature*. New York: Norton.

Cronon, William, George Miles, and Jay Gitlin. 1992. "Becoming West." in *Under an Open Sky: Rethinking America's Western Past,* edited by W. Cronon, G. Miles, and J. Gitlin. New York: Norton.

Crosby, Alfred. 1986. *Ecological Imperialism: the Biological Expansion of Europe, 900–1900*. Cambridge: Cambridge University Press.

Davenport, Walter 1935. "Power in the Wilderness." *Collier's,* September 21, 1935, pp. 10.

Davis, Arthur Powell. 1924. "Testimony Before the House Committee on Irrigation and Arid Lands." vol. y4ir7/2:c71/5–8. Washington, D.C.: Government Printing Office.

Davis, Mike. 1998. *Ecology of Fear: Los Angeles and the Imagination of Disaster.* New York: Metropolitan Books.

Descartes, Rene. 1974/1761. *Meditations on the First Philosophy.* New York: Anchor Books.

Department of the Interior. 1954–58. *Central Classified Files (Glen Canyon).* Washington D.C.

DeVoto, Bernard 1950. "Shall We Let Them Ruin Our National Parks?" *Saturday Evening Post,* 22 July 1950, pp. 17–19, 42, 44, 46, 48.

Dill, Clarence C. 1934. "Lower Cost Of Power To Help Nation." in *The Seattle Star.* Seattle, WA.

Domhoff, G. William. 1990. *The Power Elite and the State: How Policy is Made in America.* New York: Aldine De Gruyter.

Domhoff, G. William. 1996. *State Autonomy or Class Dominance? Case Studies on Policy Making in America.* New York: Aldine De Gruyter.

Dowd, M. J. 1951. *The Colorado River Flood Protection Works of Imperial Irrigation District: History, and Cost.* Imperial, CA: Imperial Irrigation District.

Dowd, M. J. 1956. *History of the Imperial Irrigation District and the Development of Imperial Valley.* Imperial, CA: Imperial Irrigation District.

Downs, Anthony. 1998. *Political Theory and Public Choice: the selected essays of Anthony Downs.* Cheltanham, UK: Edward Elger.

Eagleton, Terry. 1991. *Ideology: An Introduction.* London: Verso.

Ellison, Karin D. 1999. "The Making of a Multiple Purpose Dam: Engineering Culture, the U.S. Bureau of Reclamation, and Grand Coulee Dam, 1917–1942." Ph.D. Thesis, Program in Science, Technology, and Society, Massachusetts Institute of Technology, Cambridge, MA.

Ertman, Thomas. 1997. *Birth of the Leviathan: Building States and Regimes in Medieval and Early Modern Europe.* New York: Cambridge University Press.

Evans, Peter. 1995. *Embedded Autonomy: States and Industrial Transformation.* Princeton: Princeton University Press.

Farr, F. C. 1918. "The History of Imperial County, California." in *The History of Imperial County, California.* Berkeley: Elms and Franks.

Ficken, Robert E. 1995. *Rufus Woods, the Columbia River, and the Building of Modern Washington.* Pullman, WA: Washington State University Press.

Finegold, Kenneth and Theda Skocpol. 1995. *State and Party in America's New Deal.* Madison: University of Wisconsin Press.

Finnegan, William. 2002. "Leasing the Rain." *The New Yorker,* April 8, 2002.

Flannery, Tim Fridtjof. 1995. *The Future Eaters: An Ecological History of the Australasian Lands and People.* New York: George Braziller.

Flink, James J. 1975. *The Car Culture.* Cambridge: The MIT Press.

Flink, James J. 1988. *The Automobile Age.* Cambridge: The MIT Press.

Foucault, Michel. 1991. "Governmentality." Pp. 87–104 in *The Foucault Effect: Studies in Governmentality,* edited by G. Burchell, C. Gordon, and P. Miller. Chicago: The University of Chicago Press.

Fox, Stephen. 1981. *The American Conservation Movement: John Muir and his legacy.* Madison: University of Wisconsin Press.

Freudenburg, William R., Scott Frickel, and Robert Gramling. 1995. "Beyond the Nature/Society Divide: Learning to Think About a Mountain." *Sociological Forum* 10:361–391.

Fussell, E. B. 1938. "Coulee: Gigantic Structure Costs Millions But Benefits That Will Accrue Amply Justify Definition - 'Man's Mightiest Work.'" in *Seattle Post-Intelligencer.* Seattle, WA.

Gilbert, J. W. 1934. "City Light Head Sees East Side Domination." in *Seattle Times.* Seattle, WA.

Gilbert, Jess and Carolyn Howe. 1991. "Beyond 'State vs. Society': Theories of the State and New Deal Agricultural Policies." *American Sociological Review* 56(2): 204–20.

Glen Canyon Action Network. 2001. "Living Rivers." Moab: Glen Canyon Action Network.

Gorski, Phillip S. 2000. "The Mosaic Moment: An Early Modernist Critique of Modernist Theories of Nationalism." *American Journal of Sociology* 105:1429–1469.

Gottlieb, Robert. 1988. *A Life of Its Own: The Politics and Power of Water.* New York: Harcourt Brace Jovanovich.

Greer, F. W. 1927. *Boulder Dam and Cotton.* Held in University of Arizona Special Collections, Rollah E. Aston Collection. Columbia, SC: R. L. Bryan Co., Law Printing.

Guthrie, Woody. 1943. *Bound for Glory.* New York: New American Library.

Haber, Samuel. 1964. *Efficiency and Uplift; Scientific Management in the Progressive Era, 1890–1920.* Chicago: University of Chicago Press.

Habermas, Jurgen. 1987. *The Philosophical Discourse of Modernity: Twelve Lectures.* Translated by Frederick G. Lawrence. Cambridge: MIT Press.

González, Gustavo. 2003. "Eco-Demands Give Way to Money." http://www.ip-snews.net/interna.asp?idnews=20412. Accessed October 2003. Inter Press Service News Agency.

Guha, Ramachandra. 2000. "The Arun Shourie of the left." In *The Hindu.* Chennai (Madras).

Habermas, Jurgen. 1989. *Jurgen Habermas on Society and Politics: a reader.* Translated by Steven Seidman. Boston: Beacon Press.

Hackney, Sheldon. 1969. *Populism to Progressivism in Alabama.* Princeton: Princeton University Press.

Haraway, Donna. 1978. "Animal Sociology and a Natural Economy of the Body Politic, Part II: the past is the contested zone: human nature and theories of production and reproduction in private behavior studies." *Signs* 4.

Harvey, David. 1982. *The Limits to Capital.* Oxford: Basil Blackwell.

Harvey, David. 1989. *The Condition of Postmodernity: An Enquiry Into the Origins of Cultural Change.* Oxford: Basil Blackwell.

Harvey, David. 1996. *Justice, Nature, and the Geography of Difference.* Oxford: Blackwell Publishers.

Harvey, David. 200. *Spaces of Hope.* Berkeley: University of California Press.

Harvison, R. J. 1936. "Grand Coulee Dam Greatest Structure Built by Man." Pp. 29–31 Duboar Scrapbook, University of Washington Special Collections in *Seattle Times.* Seattle.

Hays, Samuel P. 1969. *Conservation and the Gospel of Efficiency; the Progressive Era Conservation Movement, 1890–1920.* New York: Atheneum.

Heiner, Albert P. 1989. *Henry J. Kaiser: American Empire Builder, An Insider's View.* New York: Peter Lang.

Hertzberg, Hendrik. 2001. "Changing Times." *The New Yorker,* April 23 & 30, 2001.

Hess, James. 1996. "Personal Communication." Interview conducted via electronic mail.

Hobbes, Thomas. 1982/1660. *Leviathon.* New York: Viking Press.

Holli, Melvin G. 1969. *Reform in Detroit: Hazen S. Pingree and Urban Politics.* Oxford: Oxford University Press.

Holt, L. M. 1907. "How the Reclamation Service is Robbing the Settler." *Overland Monthly* 50:510–12.

Holt, L. M. 1908. "The Reclamation Service and the Imperial Valley." *Overland Monthly* 51:70–75.

Horkheimer, Max and Theodor W. Adorno. 1944/1993. *Dialectic of Enlightenment.* New York: Continuum.

House of Representatives. 1932. *House Committee on Irrigation and Reclamation Hearings on the Columbia Basin Project, 72nd Congress, May 25, 27, June 1–3, 13, 1932,* vol. y4ir7/2:c72/4. Washington, D.C.: Government Printing Office.

House of Representatives Committee on Irrigation and Reclamation. 1924c. *Protection and Development of Lower Colorado River Basin Hearings in connection with House Resolution 2903 by Mr. Swing,* vol. y4 Ir 7/2: C71/1. Washington, D.C.: Government Printing Office.

House of Representatives Committee on Irrigation and Reclamation. 1924a. *Protection and Development of Lower Colorado River Basin Hearings in connection with House Resolution 2903 by Mr. Swing,* vol. y4 Ir 7/2: C71/5–8. Washington, D.C.: Government Printing Office.

House of Representatives Committee on Irrigation and Reclamation. 1924b. *Protection and Development of Lower Colorado River Basin Hearings in connection with House Resolution 2903 by Mr. Swing,* vol. y4 Ir 7/2: C71/4. Washington, D.C.: Government Printing Office.

Hume, David. 1974/1748. *An Enquiry Concerning Human Understanding.* New York: Anchor Books.

Hume, David. 1988/1751. *An Enquiry Concerning the Principles of Morals.* Indianapolis: Hackett.

Hundley, Norris, JR. 1972. "Clio Nods: *Arizona v. California* and the Boulder Canyon Act—A Reassessment." *The Western Historical Quarterly* 3:17–51.

Hundley, Norris, JR. 1973. "The Politics of Reclamation: California, the Federal Government, and the Origins of the Boulder Canyon Act-A Second Look." *California Historical Society Quarterly* 52:292—325.

Hundley, Norris, JR. 1975. *Water and the West.* Berkeley: University of California Press.

Hundley, Norris, JR. 1992. *The Great Thirst.* Berkeley: University of California Press.

Hunt, George W. P. 1927. "The Arizona Position." *Community Builder.*

Hunt, Lynn. 1984. *Politics, Culture, and Class in the French Revolution.* Berkeley: University of California Press.

Ickes, Harold L. 1933–41. "The Papers of Harold Ickes." Washington, D.C.: Library of Congress.

Ickes, Harold L. 1935. "Public Works for Social Gain." in *New York Times.* New York.

Ickes, Harold L. Unpublished Manuscript. *Rivers of Strength in War and Peace.* Washington, D.C.: Library of Congress, Harold Ickes Papers.

Imperial Irrigation District. 1926. "Boulder Dam All-American Canal Project and Imperial Valley, Pictorially." Imperial: Imperial Irrigation District.

Inglehart, Ronald. 1977. *The Silent Revolution: Changing Values and Political Styles among Western Publics*. Princeton: Princeton University Press.
International Rivers Network. 2003a. "Bio Bio Campaign." http://www.irn.org/programs/biobio/. Accessed October 2003: International Rivers Network.
International Rivers Network. 2003b. "Three Gorges Campaign." http://www.irn.org/programs/threeg/. Accessed October 2003: International Rivers Network.
Jackson, Kenneth T. 1985. *Crabgrass Frontier: The Suburbanization of the United States*. Oxford: Oxford University Press.
Jacobson, Paul R. 2002. *Industrialized Nature: Brute Force Technology and the Transformation of the Natural World*. Washington: Island Press.
James, Paul. 2000. "Personal Communication."
John, Candace. 2000. "Personal Interview."
Johnson, Robert Underwood. 1968. "Aesthetics and Conservation." Pp. 68–70 in *The American Environment: Readings in the History of Conservation*, edited by R. Nash. Reading, MA: Addison-Wesley Publishing Company.
Joseph, Gene. 2000. "Personal Communication."
Josephson, Paul R. 2002. *Industrialized Nature: Brute Force Technology and the Transformation of the Natural World*. Washington D.C.: Island Press.
Kahrl, William. 1982. *Water and Power*. Berkeley: University of California Press.
Kennan, George. 1917. *The Salton Sea*. New York: MacMillan and Company.
Kinsey, Don. 1928. *The River of Destiny*. Los Angeles: Department of Water and Power.
Klein, Joe. 1980. *Woody Guthrie: A Life*. New York: Alfred A. Knopf.
Kuletz, Valerie. 1998. *The Tainted Desert*. New York: Rutledge.
LaFranchi, Howard. 1998. "Indians, Environmentalists Vow Rough Water for Chile Dam." in *The Christian Science Monitor*. Boston.
Lang, William L. 1992. "A Columbia River Reader." Vancouver, WA: Center for Columbia River History.
Lears, T. J. Jackson. 1981. *No Place of Grace: Antimodernism and the Transformation of American Culture 1880–1920*. New York: Pantheon.
Lee, Lawrence. 1980. *Reclaiming the American West: An Historiography and Guide*. Santa Barbara: ABC-Clio.
Leopold, Aldo. 1991. *The River of the Mother of God and Other Essays*. Madison: University of Wisconsin Press.
Liebniz, Gottfried Wilhelm Freiherr von. 1974/1714. *The Monadology*. New York: Anchor Books.
Link, Arthur S. and Richard L. McCormick. 1983. *Progressivism*. Arlington Heights, IL: Harlan Davidson, Inc.
Litton, Martin. 1992. "Grand Canyon River Guides Oral History Project: Martin Litton Interview."
Locke, John. 1974/1690. *An Essay Concerning Human Understanding*. New York: Anchor Books
Logan, John R. and Harvey L. Molotch. 1987. *Urban Fortunes: The Political Economy of Space*. Berkeley: University of California Press.
Lotchin, Roger. 1992. *Fortress California 1910–1961, From Warfare to Welfare*. New York: Oxford University Press.

Lowitt, Richard. 1993. *The New Deal and the West.* Norman, OK: University of Oklahoma Press.

Lutz, Catherine A. and Jane L. Collins. 1993. *Reading National Geographic.* Chicago: University of Chicago Press.

Mandelbaum, Maurice. 1971. *History, Man, and Reason: A Study in Nineteenth Century Thought.* Baltimore: Johns Hopkins University Press.

Mann, Michael. 1990. *The Sources of Social Power,* vol. II: The Rise of Classes and Nation-States, 1760–1914. Cambridge: Cambridge University Press.

Manuel, Frank E. 1962. *The Prophets of Paris.* Cambridge: Harvard University Press.

Marsh, George Perkins. 1970/1864. *Man and Nature: or, Physical Geography as Modified by Human Action.* New York: Arno.

Martin, Russell. 1989. *A Story that Stands Like a Dam: Glen Canyon and the Struggle for the Soul of the West.* New York: Henry Holt and Company.

Martinez-Alier, Juan. 1995. "Political Ecology, Distributional Conflicts, Economic Incommensurability." *New Left Review* 211 (May/June).

Marx, Karl. 1867/1967. *Capital: A Critique of Political Economy, Volume I: The Process of Capitalist Production.* New York: International Publishers.

Marx, Karl. 1932/1972. "The German Ideology." in *The Marx-Engels Reader,* edited by R. C. Tucker. New York: Norton.

McCartney, Laton. 1988. *Friends in High Places, The Bechtel Story: The Most Secret Corporation and How it Engineered the World.* New York: Simon and Schuster.

McClellan, L. N. 1957. "Glen Canyon Dam." *Western Construction,* pp. 28–34.

McCluskey, H. S. 1927. "Speech to the University Club of Pasadena." Tuscon: University of Arizona Special Collections.

McCormick, Richard L. 1986. *The Party Period and Public Policy: American Politics From the Age of Jefferson to the Progressive Era.* Oxford: Oxford University Press.

McKibben, Bill. 1990. *The End of Nature.* New York: Penguin Books.

McPhee, John. 1971. *Encounters with the Archdruid.* New York: Noonday Press.

McWilliams, Carey. 1949. *California: The Great Exception.* Westport: Greenwood Press.

Mead, Elwood. 1903/1972. *Irrigation Institutions.* New York: Arno Press.

Mead, Elwood 1929a. "Conquering the Colorado." *Review of Reviews,* pp. 54–60.

Mead, Elwood. 1929b. "Memorandum for the Press: Boulder Dam." Denver: National Archives and Records Administration, Rocky Mountain Division.

Melucci, Alberto. 1996. *Challenging Codes: Collective Action in the Information Age.* Cambridge: Cambridge University Press.

Merchant, Carolyn. 1989. *Ecological Revolutions: Nature, Gender, and Science in New England.* Chapel Hill: The University of North Carolina Press.

Mills, C. Wright. 1951. *White Collar: The American Middle Classes.* Oxford: Oxford University Press.

Mills, C. Wright. 1956. *The Power Elite.* Oxford: Oxford University Press.

Moeller, Beverly Bowen. 1971. *Phil Swing and Boulder Dam.* Berkeley: University of California Press.

Morse, Kathy. 1999. "Commentary." Electronic mail discussion list.

Mould, Bob. 1992. "Hoover Dam." Minneapolis: *Copper Blue.*
Muir, John. 1908. "Letter to Sierra Club Members." Berkeley: Bancroft Library, Sierra Club Collection.
Mukerji, Chandra. 1997. *Territorial Ambitions and the Gardens of Versailles.* Cambridge: Cambridge University Press.
Mulholland, William 1927. "Water From the Colorado." *Community Builder.*
Murbarger, Nell 1957. "Dam in Glen Canyon." *Desert Magazine,* pp. 4–11.
Murlin, Bill. 1991. "Woody Guthrie, Roll On Columbia: The Columbia River Collection." Bethlehem, Pennsylvania: Sing Out Publications.
Nash, Roderick. 1967/1982. *Wilderness and the American Mind.* New Haven: Yale University Press.
National Parks Association. 1954. "Dinosaur Dams Again." *National Parks Magazine* 28:3–5.
National Utility Association. No date. "To Utility Executives." Washington, D.C.: Joint Committee of the National Utility Association.
Nelson Cook and Company. 1927. "Los Angeles and the Boulder Dam Project." Baltimore: Nelson, Cook and Company.
Neuberger, Richard L. 1938. *Our Promised Land.* New York: Macmillan.
Nixon, Edgar B. 1972. *Franklin D. Roosevelt and Conservation, 1911–1937,* vol. I, 1911–1937. New York: Arno Press.
Nye, David A. 1994. *American Technological Sublime.* Cambridge: MIT Press.
O'Connor, James. 1973. *The Fiscal Crisis of the State.* New York: St. Martin's Press.
O'Connor, James. 1998. *Natural Causes: Essays in Ecological Marxism.* New York: Guilford.
Orr, David. 1992. *Ecological Literacy: Education and the Transition to a Postmodern World.* Albany: State University of New York Press.
O'Sullivan, James. 1932. "Letter to Secretary of Agriculture Hyde." National Archives, Central Classified Documents.
O'Sullivan, James. 1934a. "Coulee Section Always Has Had U. S. Attention." in *The Seattle Star.* Seattle, WA.
O'Sullivan, James. 1934b. "O'Sullivan Hits Ross' Attack on Coulee Dam." in *Seattle Times.* Seattle, WA.
Owen, A. L. Riesch. 1983. *Conservation Under F.D.R.* New York: Praeger.
Pettitt, George A. 1935. *So Boulder Dam was Built.* Berkeley: Lederer, Street and Zeus Co.
People's Republic of China United States Embassy. 2003. "The Three Gorges Project." http://www.china-embassy.org/eng/c2718.html. Accessed 2003: The People's Republic of China.
Pinchot, Gifford. 1968. "Ends and Means." Pp. 58–64 in *The American Environment: Readings in the History of Conservation,* edited by R. Nash. Reading, MA: Addison-Wesley Publishing Company.
Pinker, Stephen. 1994. *The Language Instinct: How the Mind Creates Language.* New York: Harper Perrenial.
Pisani, Donald. 1983. "Reclamation and Social Engineering in the Progressive Era." *Agricultural History* 57:46–63.
Pisani, Donald. 1984. *From the Family Farm to Agribusiness.* Berkeley: University of California Press.

Pisani, Donald. 1992. *To Reclaim a Divided West: Water, Law, and Public Policy 1848–1902*. Albuquerque: University of New Mexico Press.

Pitkin, Hanna F. 1999. *Fortune Is a Woman: Gender and Politics in the Thought of Niccolo Machiavelli*. Chicago: University of Chicago Press.

Pitzer, Paul C. 1994. *Grand Coulee: Harnessing a Dream*. Pullman, WA: Washington State University Press.

Pollan, Michael. 1991. *Second Nature: a Gardener's Education*. London: Bloomsbury.

Pollan, Michael. 2001. *The Botany of Desire: A Plant's-eye View of the World*. New York: Random House.

Porter, Eliot. 1963. *The Place No One Knew*, Edited by D. Brower. San Francisco: Sierra Club.

Post-Intelligencer, Seattle. 1934. "Basin Project Construction is Victory for Pioneers of State." in *Seattle Post-Intelligencer*. Seattle, WA.

Powell, John Wesley. 1878. "Report of the Lands of the Arid Region of the United States." in *45 Congress, 2nd session. House Executive Document 73. Serial No. 1805*. Washington, D.C.: Government Printing Office.

Probe International. 2003. "Three Gorges Probe News Service." http://www.probeinternational.org/pi/3g/index.cfm. Accessed October 2003. Probe International.

Pyne, Stephen J. 1990. "Firestick History." *Journal of American History* 76:1132–1141.

Reisner, Marc. 1986. *Cadillac Desert*. New York: Viking.

Richardson, Elmo. 1959. "The Struggle for the Valley: California's Hetch Hetchy Controversy, 1905–1913." *California Historical Society Quarterly* 38:249–58.

Robbins, Roy M. 1976. *Our Landed Heritage: The Public Domain, 1776–1970*. Lincoln: University of Nebraska Press.

Rockwood, Charles. 1918. "The early history of Imperial County." in *The History of Imperial County*, edited by F. C. Farr. Berkeley: Elms and Frank.

Rodgers, Daniel T. 1982. "In Search of Progressivism." in *Reviews in American History*. Baltimore: Johns Hopkins University Press.

Rodgers, Daniel T. 1998. *Atlantic Crossings: Social Politics in a Progressive Age*. Cambridge: Harvard University Press.

Rorty, James 1935. "Grand Coulee." *The Nation*, March 20, 1935, pp. 329–31.

Rousseau, Jean-Jacques. 1988/1761. *The Social Contract*. Buffalo, NY: Prometheus Books.

Roy, Arundhati. 1999. *The Cost of Living*. Toronto: Vintage Books of Canada.

Sack, Robert David. 1987. *Human Territoriality and Space*. Worcester, MA: The Graduate School of Geography, Clark University.

Sanders, Elizabeth. 1999. *Roots of Reform: Farmers, Workers, and the American State, 1877–1917*. Chicago: University of Chicago Press.

Sardar Sarovar Narmada Nigam Ltd. 2003. "Narmada for People and Environment." http://www.sardarsarovardam.org/default.htm. Accessed October 2003: Sardar Sarovar Narmada Nigam Ltd.

Scattergood, E. F. 1927. "Power Possibilities and Costs." *Community Builder*.

Schaefle, William J. 1928. "The People of the United States Do Not Want Another Tea Pot Dome Scandal." *American Globe* 25:5–7.

Schnaiberg, Allan. 1980. *The Environment: From Surplus to Scarcity.* New York: Oxford University Press.
Schonfeld, Robert G. 1969. *The Early Development of California's Imperial Valley.* Los Angeles: The Historical Society of Southern California.
Scott, James. Unpublished Manuscript. *The TVA.*
Scott, James C. 1998. *Seeing Like a State: How Certain Schemes to Improve the Human Condition Have Failed.* New Haven: Yale University Press.
Seliger, Martin. 1976. *Ideology and Politics.* London: Allen and Unwin.
Senate. 1926. *Senate Hearings from the Committee on Irrigation and Reclamation,* vol. y4.ir7/a:c72/2. Washington, D.C.: United States Senate.
Senate. 1928. *Congressional Record, 70th Congress 1st Session,* vol. 69, part 9. Washington: Government Printing Office.
Sewell, William H., JR. 1980. *Work and Revolution in France: The Language of Labor from the Old Regime to 1848.* Cambridge: Cambridge University Press.
Shallat, Todd. 1995. *Structures in the Stream: Water, Science, and the Rise of the U.S. Army Corps of Engineers.* Austin: University of Texas Press.
Sierra Club. 1911. "To the Members of the Sierra Club and Their Friends." Berkeley: Bancroft Library, Sierra Club Collection.
Sierra Club. 1954. "Two Wasteful Dams—Or A Great National Park?" *The Sierra Club Bulletin* 39:3–4.
Sierra Club. 1957. *Public Relations Primer.* San Francisco: Sierra Club.
Simon, Julian. 1981. *The Ultimate Resource.* Princeton: Princeton University Press.
Skocpol, Theda. 1979. *States and Social Revolutions: a Comparative Analysis of France, Russia, and China.* New York: Cambridge University Press.
Skocpol, Theda. 1985. "Bringing the State Back In: Strategies of Analysis in Current Research." in *Bringing the State Back In,* edited by P. B. Evans, D. Rueschemeyer, and T. Skocpol. New York: Cambridge University Press.
Skocpol, Theda. 1992. *Protecting Soldiers and Mothers: The Political Origins of Social Policy in the United States.* Cambridge: Harvard University Press.
Skowronek, Stephen. 1982. *Building a New American State; the Expansion of National Administrative Capacities, 1877–1920.* Cambridge: Cambridge University Press.
Smith, Neil. 1984. *Uneven Development: Nature, Capital and the Production of Space.* Oxford: Basil Blackwell.
Special Advisors to the Secretary of the Interior, Bureau of Reclamation. 1928. *Development of the Lower Colorado River.* Washington, D.C.: Government Printing Office.
Spokane Chamber of Commerce. 1936a. "Grand Coulee Dam." Spokane: Spokane Chamber of Commerce.
Spokane Chamber of Commerce. 1936b. "Grand Coulee Dam and its Headquarters Spokane." Spokane, WA: Spokane Chamber of Commerce.
Spokane Chamber of Commerce. 1938. *The Columbia Basin Grand Coulee Project: A Remarkable National Resource that will Contribute Perpetually to the Country's Wealth, Prosperity, and Well-Being.* Spokane, WA: Spokane Chamber of Commerce.
Squires, Charles. 1928. "Dam Bill is Before Congress." in *Las Vegas Age.* Las Vegas.

Squires, Charles P. 1865–1958. "Squires Papers." Las Vegas: University of Nevada Special Collections.

Staff 1935. "Boulder Dam Harnesses Mighty Colorado to Make It work for Mankind." *The Pathfinder,* March 23, 1935.

Star, The Seattle. 1934. "Immensity of Grand Coulee Project Hard to Appreciate; Act of Harnessing Tremendous Power of Columbia is Hailed as Record Feat of Engineering." in *The Seattle Star.* Seattle, WA.

Stegner, Wallace. 1955. *This is Dinosaur: Echo Park Country and Its Magic Rivers.* New York: Alfred A. Knopf.

Stegner, Wallace. 1969. "Glen Canyon Submersus." in *The Sound of Mountain Water.* Garden City: Doubleday and Company.

Stegner, Wallace E. 1954. *Beyond the Hundredth Meridian: John Wesley Powell and the Second Opening of the West.* Boston: Houghton, Mifflin.

Steinmetz, George. 1999. "Introduction." in *State/Culture: State-Formation after the Cultural Turn,* edited by G. Steinmetz. Ithaca: Cornell University Press.

Stevens, Joseph E. 1988. *Hoover Dam: An American Adventure.* Norman, OK: University of Oklahoma Press.

Stratton, Owen and Phillip Sorotkin. 1955. "The Echo Park Controversy and Upper Colorado River Development." in *Department of Interior Central Classified Files box 363.* College Park: National Archives and Records Administration.

Stuart, Gordon. 1961. *When the Sands of the Desert Grew Gold:* Self-published by the author.

Sundborg, George. 1954. *Hail Columbia: The Thirty-Year Struggle for Grand Coulee Dam.* New York: MacMillan.

Swain, Donald C. 1963. *Federal Conservation Policy, 1921–1933.* Berkeley: University of California Press.

Swing, Phil D. 1890–1963. "Phil D. Swing Papers." Los Angeles: UCLA Special Collections.

Teale, Edwin Way. 1954. *The Wilderness World of John Muir.* Boston: Houghton Mifflin Company.

The Colorado River Land Company. 1924. *The Other Side of the Question: A Statement by The Colorado River Land Company.* Los Angeles: The Colorado River Land Company.

The Colorado River Land Company. 1926. "The Federal Government's Colorado River Project." Los Angeles: The Colorado River Land Company.

Thelen, David P. 1976. *Robert M. LaFollette and the Insurgent Spirit.* Boston: Little, Brown.

Thompson, Hunter S. 1971. *Fear and Loathing in Las Vegas: A savage journey into the heart of the American dream.* New York: Vintage.

Thompson, John. 1990. *Ideology and Modern Culture.* Stanford: Stanford University Press.

Thomas, John L. 2000. *A Country in the Mind: Wallace Stegner, Bernard DeVoto, History, and the American Life.* New York: Routledge.

Tilly, Charles. 1990. *Coercion, Capital, and European States, AD 990–1990.* Cambridge, MA: Basil Blackwell.

De Tocqueville, Alexis. 1840/1990. *Democracy in America Volume II.* New York: Vintage Classics.

Turner, Christy G. and Jacqueline A. Turner. 1990. "Perimortem Damage to Human Skeletal Remains from Wupatki National Monument, Northern Arizona." *Kiva* 55:187–209.

Turner, Frederick Jackson. 1920/1996. *The Frontier in American History.* New York: Dover Publications.

Union-Bulletin, Walla Walla Sunday. 1934. "President Recalls his visit of 1920 to Columbia Basin; Roosevelt thought then that river should be developed; full text of speech given." in *Walla Walla Sunday Union-Bulletin.* Walla Walla, WA.

United States Bureau of Reclamation. 1996. "U.S. Bureau of Reclamation Home Page." http://www/usbr.gov. Accessed 2000. Department of the Interior.

Vandergeest, Peter. 1996. "Mapping Nature: Territorialization of Forest Rights in Thailand." *Society and Natural Resources* 9:159–175.

Vandergeest, Peter and Nancy Lee Peluso. 1995. "Territorialization and state power in Thailand." *Theory and Society* 24:385–426.

Wallerstein, Immanuel. 1974. "The Rise and Future Demise of the Capitalist World System: Concepts for Comparative Analysis." *Comparative Studies in Society and History* 16:387–415.

Walton, John. 1992. *Western Times and Water Wars.* Berkeley: University of California Press.

Water Resources. 1928. "Water Resources." vol. IV. New York: Water Resources Publishing Company.

Webb, Walter Prescott. 1931/1959. *The Great Plains.* New York: Ginn and Company.

Weber, Max. 1904–19/1949. *Methodology of the Social Sciences.* New York: The Free Press.

Weber, Max. 1905/1958. *The Protestant Ethic and the Spirit of Capitalism.* New York: Charles Scribner's Sons.

Weber, Max. 1909/1976. *The Agrarian Sociology of the Ancient Civilizations.* London: R. I. Frank.

Weber, Max. 1922/1946. "Science as a Vocation." Pp. 129–156 in *From Max Weber,* edited by H. Gerth and C. W. Mills. New York: Oxford University Press.

Weber, Max. 1922/1978. *Economy and Society.* Berkeley: University of California Press.

Weber, Max. 1923/1961. *General Economic History.* New York: Collier.

Weber, Max. 1946. *From Max Weber: Essays in Sociology.* Translated by H. H. Gerth and C. Wright Mills. Oxford: Oxford University Press.

Weber, Max. 1952. *Ancient Judaism.* Translated by Hans Gerth and Don Martindale. New York: The Free Press.

Weil, Fred M. 1938. *A Curb-Side Story of Grand Coulee Dam and a Last Frontier.* Grand Coulee, WA: Teepee Information Services.

Wheeler, John R. 1934. "Projects Along Columbia To Be World Greatest." in *The Seattle Star.* Seattle, WA.

White, Owen P. 1934. "Spare That Desert!" *Collier's,* June 16, 1934, pp. 10.

White, Richard. 1980. *Land Use, Environment, and Social Change: The Shaping of Island County, Washington.* Seattle: University of Washington Press.

White, Richard. 1983. *The Roots of Dependency: Subsistence, Environment, and Social Change Among the Choctaws, Pawnees, and Navajos.* Lincoln: University of Nebraska Press.

White, Richard. 1991. *"It's Your Misfortune and None of My Own;" a History of the American West.* Norman, OK: University of Oklahoma Press.
White, Richard. 1995. *The Organic Machine.* New York: Hill and Wang.
Wiebe, Robert H. 1967. *The Search for Order, 1877–1920.* New York: Hill and Wang.
Wilderness Society. 1954. "Slick Paper with Slick Words." *The Living Wilderness* 19:27–30.
Williams, Albert N. 1951. *The Water and the Power: Development of the Five Great Rivers of the West.* New York: Duell, Sloan, and Pearce.
Williams, Raymond. 1976. *Keywords: A Vocabulary of Culture and Society.* Glasgow: Fontana.
Williams, Raymond. 1980. *Problems in Materialism and Culture.* London: NLB.
Wilson, Richard G. 1985. "Machine-Age Iconography in the American West: The Design of Hoover Dam." *Pacific Historical Review* 54:463–493.
Wilson, Richard G. 1986. *The Machine Age in America, 1918–1941.* New York: Harry N. Abrams, Inc.
Wirth, Conrad. 1980. *Parks, Politics, and the People.* Norman: University of Oklahoma Press.
Wisconsin Stewardship Network. 2003. http://www.wsn.org/shorelands/dams.html. Accessed 2003.
Wittfogel, Karl A. 1957. *Oriental Despotism.* New Haven: Yale University Press.
Woodbury, David O. 1941. *The Colorado Conquest.* New York: Dodd, Mead and Company.
Woods, Rufus. 1918. "FORMULATE BRAND NEW IDEA FOR IRRIGATION GRANT, ADAMS, FRANKLIN COUNTIES, COVERING MILLION ACRES OR MORE." pp. 7 in *Wenatchee Daily World.* Wenatchee, WA.
Woods, Rufus. 1937a. "The Cavalcade of America." in *The Eighth Wonder of the World,* edited by WABC: DuPont.
Woods, Rufus. 1937b. "Special Edition." in *Wenatchee Daily World.* Wenatchee, WA.
Woods, Rufus. 1944. *The 23-Years' Battle for Grand Coulee Dam.* Wenatchee, WA: Wenatchee Daily World.
Work, Hubert 1927. "The Importance of Developing the Lower Colorado." *The Congressional Digest,* February, pp. 39–40.
Work, Hubert and Elwood Mead. 1926. *Studying Reclamation.* Washington: Department of Interior.
World Commission on Environment and Development. 1987. *Our Common Future.* Oxford: Oxford University Press.
Worster, Donald. 1985. *Rivers of Empire: Aridity and the Growth of the American West.* Oxford: Oxford University Press.
Worster, Donald. 1990. *Under Western Skies.* Oxford: Oxford University Press.
Worster, Donald. 1994. *Nature's Economy: A History of Ecological Ideas.* Second Edition. Cambridge: Cambridge University Press.
Young, C. C. 1927. "The California Position." *Community Builder.*

Index

A

Abbey, Edward; *see also* Dams, Glen Canyon, opposition; 213–216
Adams, Ansel, 197
Agriculture; *see also* Irrigation; 20, 31, 43, 52, 74, 120, 172, 215
All American Canal
 Boulder Dam and, 29, 76–83
 discourse of national importance and, 88, 91, 92
 Ed Sample and, 121
 irrigation and, 78, 82
 Mexico and, 76, 78, 112
 nature and, 77
 National Utility Association and, 123, 125
 L.A. Times and, 115, 118–120, 123, 132, 247
American Legion, 87, 91–92, 103
Antimodernism, 18
Arid land
 Bureau of Reclamation and, 25, 30
 environmentalism and, 191, 236
 state building and, 7, 30, 188, 218
 water rights and, 44
Arizona
 Boulder Dam opposition and, 116, 125, 127–132
 Colorado River Compact and, 62, 117
 congressional delegation of, 103, 107
 Glen Canyon Dam and, 190, 194
Army Corps of Engineers
 Bureau of Reclamation and, 3, 7, 143, 168, 171, 209
 dam construction generally and, 3, 7
 Grand Coulee and, 137, 143,168, 171
 opposition to dams and, 196
 National Park Service and, 209
Automobile; *see also* Interstate Highway System; 38, 228

B

Biobio River, 230
Boulder Dam and, 117, 245–246
 conception of nature at, 66, 69
 images of, 113
 named, 245–246
 natural features of, 66, 69, 112, 113
Bonneville Power Administration, 140, 178
Boulder Canyon
 Act, 83–86, 108, 123–124, 130, 246
 map of, 239
 named, 45, 52, 55, 245–246
 Project, 55, 62, 79, 86–7, 106, 111
Boulder Dam, *see* Dam, Boulder
Boulder Dam Association, 87–91, 93, 102
Brower, David
 Dinosaur National Monument and, 55, 197–200, 210–211
 Glen Canyon Dam and, 211–216
 Grand Canyon and, 217
 Rainbow Bridge and, 216
 Sierra Club and, 197–200, 211
Bureau of Indian Affairs, 160, 215, 218
Bureau of Reclamation
 All American Canal and, 76
 Boulder Dam and, 77
 cooptation of discourse and, 217–220
 Dinosaur National Monument and, 196
 discourse and, 4, 5, 64, 69, 87, 108–115, 171–175, 188–189, 191–195, 236–237
 Glen Canyon Dam and, 188–189, 191–195

Bureau of Reclamation *(continued)*
- Grand Coulee Dam and, 139, 154, 171–174
- imperialistic discourse and, 87, 97, 108–115, 171, 174, 188, 225
- Native Americans and, 215–216, 218
- periods of reclamation and, 54–56
- as a Progressive-Era institution, 42, 52
- state simplifications and, 25
- Grange (The) and, 158
- National Park Service and, 204–207, 209
- views of nature and, 69, 111

C

California
- agricultural economy of, 43, 76, 120
- Arizona and, 127–130
- Colorado River Compact and, 62
- Doctrine, 44
- economy of, 42
- water policy and, 43–45, 75
- water projects in, 56, 235
- water scarcity and, 42
- and the West, 55, 62

Capital
- accumulation of, 7, 36, 39–40, 229
- control of space and, 34, 35
- private, 3, 4, 6, 30, 34, 52, 75, 126, 229, 230, 244
- production of nature and, 33–35
- treadmill of production and, 18–20, 244
- workers/labor and, 18–20, 71, 100

Car, *see* automobile

Chaffey, George, 73

Chandler, Harry, 90, 115–123, 131, 247

Chile, 230

China, 174, 231–232

Civil society
- Boulder Dam and, 87, 96, 103
- Grand Coulee Dam and, 134, 183, 189, 227
- Glen Canyon Dam and, 195, 226, 228
- the state and, 10, 18, 28, 39–40, 229
- anemic/weak, 25–27, 233
- strong, 26–27, 36, 87, 96, 229

Clapp, William, 138, 142, 143

Class, 13, 19, 20, 22–23, 26–27, 28, 37–38, 40, 188, 197, 200, 228, 233, 245

Colville Reservation, *see* Confederated Tribes of the Colville Reservation

Colorado River
- Boulder Dam and, 3, 62
- discursive construction of, 89, 223
- flow of, 55, 79, 117, 128
- Glen Canyon Dam and, 4, 190, 191
- as "menace," 16, 60, 68, 80, 192–194
- as "miracle," 188
- as "resource," 60, 64, 72, 77

Colorado River Compact of 1922
- Arizona and, 103, 116, 117, 127
- Boulder Dam and, 45, 55, 62,
- Herbert Hoover and, 97, 98, 246

Colorado (State of), 191, 203

Columbia Basin Commission, 158

Columbia Basin Irrigation League, 155, 159

Columbia Basin project, 148–155

Columbia River
- Anglo discovery of, 137
- Bonneville Dam and, 63
- discursive construction of, 154, 156, 170, 174, 184
- flow of, 136
- Grand Coulee and, 3, 63, 137
- Woody Guthrie and, 6, 136, 178–183

Columbia River Development League, 148

Confederated Tribes of the Colville Reservation, 159–161

Conservation
- Harold L. Ickes and, 71–72, 100, 177
- national defense and, 157
- New Deal and, 52–54
- Progressive Era and, 45–52 188

Contestation
- Boulder Dam and, 114, 115, 131, 162
- Glen Canyon and, 189, 215
- Grand Coulee and, 162, 181, 183, 188
- high modernism and, 5
- ideology and, 13, 235
- over dams, 3–4

Coolidge, Calvin, 51, 86, 97–98, 126

Cooperation (with nature); *see also* Harold L. Ickes; 71–72, 100–101, 106, 133

Cooptation, 190, 224, 237, 238

Crisis
- agricultural, 139
- capital and, 4, 14, 35
- ecological, 14
- fiscal, 19
- legitimation, 19
- water, 1, 234, 235

Culture
  frontier, 23, 28, 75
  legitimation and, 228
  material, 29, 229
  nature and, 32
  political, 26, 27
  popular, 133

D

Dams
  Boulder
    construction of, 61–63
    discourse at, 4, 7, 16, 51, 60, 63–73
    Ickes, Harold L. and, 71–72, 100
    machine aesthetic and, 133–134
    Native Americans and, 215
    opposition to, 37, 60, 115–132
    PWA and, 53, 99
    Roosevelt, Franklin D. and, 71, 97–102, 246
    second phase of reclamation and, 54–55
  Bonneville, 63, 100, 137–140, 163, 170–171, 178
  Echo Park; *see also* Dams, Glen Canyon; 4, 55, 188–224
  Fort Peck, 133
  Glen Canyon
    consequences of, 228, 229, 234–238
    construction of, 190–191
    discourse at, 4, 7
    opposition to, 188–189, 195–210
    Native Americans and, 215–216, 217–219
    third phase of reclamation and, 54–55
  Grand Coulee, 7, 37, 52, 53, 61
    construction of, 136–140
    discourse at, 4, 7, 51, 63, 135, 140–143
    Harold L. Ickes and, 100, 176–178
    machine aesthetic and, 133–134
    Native Americans and, 159–162
    opposition to, 37, 164–168
    PWA and, 53, 138,164
    Woody Guthrie and, 178–183
  Hetch Hetchy, 50–51, 64, 115, 195–197, 205, 211, 223
  Hoover, *see* Boulder
  Laguna, 54, 55, 92
  Ralco, 230
  Sardar Sarovar, 230–231
  Shasta, 63, 134
  Split Mountain; *see also* Dams, Echo Park; Dams, Glen Canyon; 191–192, 195–196, 206, 210–211
  Three Gorges, 231–232
Dam removal, 217, 246
Davis, Arthur Powell (Commissioner of Reclamation), 20, 54–55, 64, 69, 76–79, 83–84, 105, 108–110, 172
Department of Interior, 78, 159, 171, 178, 200, 206, 207, 208, 215
Development
  agricultural, 145, 169, 232
  blind, 205, 237
  economic, 18–20, 35, 42, 49, 111, 120, 133, 145, 170, 203, 224, 227, 229
  industrial, 4, 18–20, 107, 111, 145, 169, 227
  infrastructural, 22, 73, 78, 97, 232
  social, 20, 47, 227
  sustainable, 14
  treadmill of production and, 18–20, 229–233
DeVoto, Bernard, 37, 196–197, 199
Dialectic of nature and society, *see* Nature-society dialectic
Dinosaur National Monument; *see also* Echo Park; Glen Canyon; 37, 55, 191, 196–211, 223, 227, 237
Discourse; *see also* Ideology
  analysis of, 3, 4–5
  anti-Hebraic, 162–168
  defined, 13
  Hebraic, 7, 141, 154–162
  high modernist, 5, 114, 146, 171, 188, 218, 227
  ideology and, 11–13
  imperialistic, 78, 87, 97, 99–107, 132, 141, 144, 148, 169, 170, 175–177, 179, 181
  of improved nature, 190, 219, 223, 235, 238
  individualist-capitalist, 7, 126, 141, 151, 154, 162, 184
  language and, 11–13
  of national importance, 78, 87–97, 132, 188
  oppositional, 4, 120, 126, 132, 141, 183, 185, 188–190, 196, 209, 223–224, 237
  state-centered, 7, 94, 141

Domination
gender and, 15–16
of humans, 13, 15
of nature, 3, 4, 7, 10, 14–15, 18, 23, 28–30, 36, 42, 228–229, 233, 238
Dominy, Floyd (Commissioner of Reclamation), 191, 215, 220

E

Echo Park, *see* Dams, Echo Park
Eisenhower, Dwight D., 200, 203
Ellinwood, Thomas R., 194
Enlightenment; *see also* Modernity; Rationality; 15–17

F

Fall, Albert B. (Secretary of Interior), 78, 92, 108, 247
Fall-Davis Report, 78, 92, 108
Fiscal crisis *see* Crisis, fiscal
Flood control
Boulder Dam and, 63, 77–82, 99
Glen Canyon Dam and, 194
Grand Coulee Dam and, 137
multiple purpose river development and, 52, 63, 84, 86, 110, 141, 142
single purpose river development and, 9, 20, 53, 63, 88, 98, 111, 115, 118–122, 194, 231
French Revolution, 12, 16–17
Frontier
culture of , 36, 50, 141, 193–195
last frontier, 141, 147, 155, 166
nature and, 19, 141, 189, 194
social control and, 28, 39, 48, 162
southern frontier, 229–233
state building and, 39–43, 48, 57, 162
treadmill of production and, 19, 20
Turner's frontier thesis, 23, 26, 50
western frontier, 20, 23

G

Gender, 10, 13–15, 73, 106, 185
Glen Canyon
Dam, *see* Dams, Glen Canyon
discovered, 190–191
map of, 241
named, 190
natural features of, 190, 212–214
Grand Coulee
Dam, *see* Dam, Grand Coulee
discovered, 137
gravity plan and, 143–144
map of, 240
named, 137
natural features of, 136–137
pumping plan and, 143–144
Grange (The), 158–159
Green River, 191
Guthrie, Woody
Bonneville Dam and, 178
Columbia River and the, 178
Department of Interior and, 6, 178
Grand Coulee Dam and, 136, 178–183, 224
as social critic, 1, 39–40, 180–183, 184

H

Hanford Nuclear Reservation, 140
Harding, Warren G., 51, 97, 247
Hayden, Carl (U.S. Representative and Senator from Arizona), 103, 107, 130
Hegemony/hegemonic, 4, 7, 26, 36, 185, 188, 223–224, 232, 237
Hetch Hetchy Valley; *see also* Yosemite National Park; 50–51, 64, 115, 195, 196, 205, 211, 223
High modernism; *see also* Discourse
Boulder Dam and, 114
defined, 25, 245, 248
discourses of, 3–5, 7, 225, 226–229, 233–237
Glen Canyon Dam and, 185, 226
Grand Coulee Dam and, 145, 183–185
state building and, 25–27, 29, 36, 47
Hoover, Herbert, 51, 52, 62, 97, 98–99, 246
Hunt, W. P. (Arizona Governor), 127–129
Hyde, Arthur M. (Secretary of Agriculture), 148
Hydropower; *see also* Irrigation; Flood Control
Boulder Dam and, 83–86, 94, 101, 110, 132
consequences of, 234, 237, 248
Glen Canyon Dam and, 200, 203, 216, 220
Grand Coulee Dam and, 137, 139, 156, 158, 159–162, 176, 184
multiple purpose river development and, 77–78, 83–86, 231–234

Native Americans and, 159–162, 216
public ownership of, 115
sale of, 78, 89, 91, 159
single purpose river development and, 51, 71
technology of, 77

I

Ickes, Harold L. (Secretary of Interior)
Boulder Dam and, 71–72, 99–102
concept of cooperation, 71, 72, 99–102, 106, 133, 185
conception of nature, 71–72, 99–102, 106, 133, 177, 185
conservation and, 71, 177
Franklin D. Roosevelt and, 54, 99
Grand Coulee Dam and, 138, 143, 160, 171, 176–178
Native Americans and, 160
New Deal and, 54, 99, 171
Progressivism and, 54
Ideology; *see also* Discourse
environmentalist, 223, 237
defined, 10–13
of domination, 23, 42, 136, 142, 238
of economic growth, 19, 235, 238
hegemonic, 3, 4, 5, 7, 224, 237
of high modernism, 4, 5, 7, 25–27, 29, 141, 172, 184, 233, 235, 237, 245
of imperial modernism, 189, 193, 223, 224
machine age, 133
of modernism, 47, 133
progressive, 2, 52, 144
utilitarian, 42
Imperial Irrigation District (IID); *see also* All American Canal; Dams, Boulder; 60, 75, 82, 93, 115, 119, 121
Imperial Valley of California, 87, 92, 98, 108–120, 134, 227
Imperialist discourse *see* Discourse, imperialist
India, 230–231
Indians, *see* Native Americans
Infrastructure
compression of space and, 34
economic, 2, 22, 27, 36, 40, 49, 78, 183, 226
political, 2, 22, 183, 226
state-sponsored, 2, 26, 233
transportation, 20, 38, 39
water, 7, 26, 36, 57, 78, 224, 226, 229
Interstate Highway System, 38
Irrigation; *see also* Hydropower; Flood Control
Boulder Dam and, 62, 82–83
financing of, 47, 89, 108
Glen Canyon Dam and, 216–218
Grand Coulee Dam and, 137, 143–144
multiple purpose river development and, 2, 7, 25, 30, 33, 41, 62, 77–78, 138, 230–232
Native Americans and, 159–160, 218–220
single purpose river development and, 61, 115
social structure of, 75

J

Johnson, Hiram (U.S. Senator from California), 69, 77, 102, 105–106, 130
Johnson, Robert Underwood, 50

K

Kaiser, Henry J., 63, 134
Kaufmann, Gordon B., 134
Knopf, Alfred A., 197–200

L

Labor, 7, 18, 19–20, 24, 34, 36, 37, 48, 64, 71, 82, 100, 233, 244
Las Vegas, Nevada, 96, 237
Legitimation, 19, 64, 192, 195, 228
Leopold, Aldo, 15, 226, 238
Los Angeles, California, 62, 80, 84, 105, 120, 123–125, 130, 193
Los Angeles Times, 90, 116–123, 197

M

Material Culture, *see* Culture, material
Material power, 4, 237
McKay, Douglas (Secretary of Interior), 200, 204, 210–211
Mead, Elwood (Commissioner of Reclamation), 54–56, 108–111, 169, 225, 248
Mexico, 39, 44, 74, 75, 76, 91, 112, 115, 117, 134
Middle class, 26, 37, 40, 188, 197, 200, 245

Modernity; *see also* Antimodernism; Domination; 10, 13, 15–17, 34, 47
Muir, John, 7, 49, 50–51, 64–65, 115, 120, 132, 195, 245
Mulholland, William, 128–129
Myth of the west, 1, 28, 29, 69, 80

N

Nakai, Raymond (Navajo Nation Tribal Council Chairman), 215
Narmada River, 230, 231
National Park Service, 197, 203, 205, 206–209
National Utility Association, 116, 123–126, 132
Native Americans
  Boulder Dam and, 215
  Glen Canyon Dam and, 215–216, 217–219
  Grand Coulee Dam and, 159–162
  views of nature and, 14, 32
Natural resources; *see also* Water; *see also* Nature;
  abuse of, 14, 144, 223, 237
  aesthetic use of, 188, 195, 204, 209, 212
  control of, 20
  New Deal and, 52, 53
  rivers as, 60, 89
  social construction of, 60
  use of, 13, 14, 37, 72, 177, 223, 237
Nature; *see also* Nature-society dialectic; Wilderness
  aesthetic, 10, 49, 132, 188–190, 195, 203–205, 212–213, 222–224, 237–238, 249
  consumption of, 49, 188–189, 222
  dialectic of, 8, 33, 39–41
  domination of, 4, 14–16, 18, 36, 65–68, 101, 132, 142, 154, 185, 223, 226–229
  as economic resource, 16, 28, 29, 32, 64, 78, 87, 92, 132, 193, 194, 195, 249
  as feminine, 10, 13–15, 73, 106, 185
  "first" nature, 34
  human conceptions of, 31, 40, 236
  improved, 190, 219, 223, 235, 238, 249
  nature-society relationship, 18, 41, 106, 129, 136, 184, 189, 223, 235, 237
  production of, 34
  "second" nature, 34
  social construction of, 16, 23, 29, 35, 40, 60, 73, 78, 129, 132, 193, 195, 205, 219, 229, 235
  transformations of, 4, 34, 42, 70, 243
Nature-society dialectic; *see also* Nature; 8, 9–10, 31–40, 41
New Deal
  Boulder Dam and, 97, 99, 102
  conservation programs, 42, 52–54
  Grand Coulee and, 138, 139, 164, 172, 176
  high modernism and, 27,
  Ickes, Harold L. and, *see* Ickes, Harold L.
  land programs, 37, 52–53
  Progressivism and, 51–52
  Roosevelt, Franklin D. and, 54, 99, 138, 168, 171, 172
  social engineering and, 167, 168
  treadmill of production and, 19
  utilitarian ideology and, 2
Newell, Frederick H. (Commissioner of Reclamation), 54, 150

O

Oregon, 137, 159
O'Sullivan, James, 142, 147, 148–152, 171

P

Phoenix, Arizona, 193
Political economy, 3, 7, 13, 18, 34, 39
Porter, Eliot, 197, 212–213, 215
Powell, John Wesley, 76, 190, 199, 212, 243
Preservation
  conservation and, 49–51
  environmentalism and, 55, 188, 203, 237
  of wilderness, 38, 195, 199, 223
Progressive Era, 2, 19–20, 24, 25, 27, 42, 45–52, 90, 99, 144, 226, 233, 245
Progressivism, 42, 45–52, 54, 129, 130, 175
Public Works Administration (PWA), 53, 99, 138, 164
Pinchot, Gifford, 24, 49, 50

R

Rainbow Bridge National Monument, 216, 217
Rationalization; *see also* Modernity; 15–18, 24, 42, 47, 49

Reclamation Act of 1902; *see also* Bureau of Reclamation; 25, 47, 49, 52, 54, 62, 108
Reclamation Service, *see* Bureau of Reclamation
Reformation (The), 16
Rockwell, Norman, 38, 221, 222
Rockwood, Charles, 73–74, 112
Roosevelt, Franklin D.
Boulder Dam and, 71, 97–102, 246
  conservation and, 52–53, 71
  Grand Coulee Dam and, 138–139, 143, 151, 172, 174–175
  Ickes, Harold, L. and, 54, 71, 99, 143
  New Deal and, 54, 99, 138, 168, 171, 172
  Progressivism and, 175
  social engineering and, 167, 168
Roosevelt, Theodore, 18, 47, 62, 80, 82, 150, 175
Rose, Mark, 60, 75–76, 92
Ross, J. D., 142, 151
Roy, Ahrundhati, 231

S

Salton Sea, 74
Sample, Ed (California State Senator), 70–71, 121–123, 247
San Diego, California, 193
San Francisco, California, 50–51, 64, 196
Savage, John L., 134
Sierra Club
  Glen Canyon and, 37, 55, 197, 203–206, 210–212, 217, 225, 236
  Hetch Hetchy Dam, 49–51, 115, 196
Six Companies, 63, 65, 112, 134
Social construction of reality, 11–13
Social control, 22, 25, 28, 49
Social engineering, 26, 53, 167–168
Southern Pacific Railroad (SPRR), 74
Space, 13, 34–35, 39
Spokane Chamber of Commerce, 155–157, 162
State (The)
  building, 2, 4, 20, 21–23, 23–26
  capacity, 1, 22, 25, 39, 56, 234
  civil society and, 10, 18, 27, 28, 39–40, 229
  control, 3, 4, 20, 22, 23–30, 36, 45, 48, 78, 107, 133, 172, 224, 229
  crisis and, 14, 35, 19–20, 139, 234
  culture and, 3, 7, 10, 17, 26, 226–229
  defined, 244
  as plurality, 5, 21, 244
  power, 18, 21, 24, 32–36, 40, 229, 232
  simplifications, 23–30, 226
  territoriality and, 4, 23–29, 34–36, 48, 84, 107, 133, 229
  theories of, 21–23
  treadmill of production and, 19–21, 244
Stegner, Wallace, 37, 197–200, 236
Strauss, Michael (Commissioner of Reclamation), 191
Sustainability, 14–15, 231
Swing-Johnson Bill, *see* Boulder Canyon Act
Swing, Phil (U.S. Representative); *see also* All American Canal; Dams, Boulder; Imperial Irrigation District; 60, 69, 76–77, 82–84, 88, 102–112, 120–123, 247

T

Taylorism, 48
Teapot Dome scandal, 86, 247
Tennessee Valley Authority (TVA), 53, 203, 245, 248
Territoriality, *see* State, territoriality
Time, 13, 17, 27, 34–35, 39,
Title transfers, 56, 234
Tourism/tourist, 38, 63, 133, 147, 156, 158, 207, 216
Turner, Frederick Jackson; *see also* Frontier; 23, 26, 50
Treadmill of production; *see also* Capital; Development; 14, 18–22, 229–233

U

Udall, Stewart (Secretary of the Interior), 220
Unintended consequences, 4, 29, 34, 38, 42, 233–235
U.S. Forest Service, 24
U.S. Geological Survey, 190
U.S. Park Service, 197, 203–209
Utah, 62, 116, 125, 188, 190, 203, 207, 212
Utilitarianism; *see also* Progressivism; 24, 42, 47, 50, 51, 62, 71, 100–101, 185

W

Water; *see also* Nature; *see also* Natural resources
  control of, *see* Flood Control

Water *(continued)*
policy, 43–45
rights, 43, 44, 74
scarcity, 2, 6, 7, 20, 23, 41, 234
Water Resources Association (WRA), 126
Wilderness; *see also* Nature
Act (1964), 55
preservation of, 38, 49, 55, 210
Society, 189, 190, 200, 204, 208–209
the trouble with, 189
Wirth, Conrad (National Park Service Director), 209
Woods, Rufus, 142, 143, 145–147, 156
Work, Hubert (Secretary of Interior), 69, 104, 107–109, 121, 248
World War I, 143, 150, 243
World War II, 38, 140, 160, 222
Wyoming, 44, 45, 62, 247

Y

Yampa River, 191, 204
Yangtze River, 231
Yosemite National Park; *see also* Hetch Hetchy Valley; 50, 64, 115, 195–196

For Product Safety Concerns and Information please contact our EU
representative GPSR@taylorandfrancis.com
Taylor & Francis Verlag GmbH, Kaufingerstraße 24, 80331 München, Germany

www.ingramcontent.com/pod-product-compliance
Ingram Content Group UK Ltd.
Pitfield, Milton Keynes, MK11 3LW, UK
UKHW022117190926
475702UK00013B/1488
* 9 7 8 0 4 1 5 6 4 5 8 0 5 *